Knowledge as Commons

Knowledge as Commons

Toward Inclusive Science and Technology

PRABIR PURKAYASTHA

MONTHLY REVIEW PRESS
New York

Copyright © 2024 by Monthly Review Press
All Rights Reserved

Originally published in India by LeftWord Books © LeftWord Books 2023.
U.S. edition published by Monthly Review Press 2024.

Library of Congress Cataloging-in-Publication Data
available from the publisher

ISBN: 978-1-68590-070-0 paper
ISBN: 978-1-68590-071-7 cloth

MONTHLY REVIEW PRESS, NEW YORK
monthlyreview.org

5 4 3 2 1

For
Amit Sengupta,
Aijaz Ahmad,
A. Gopalakrishnan
and
the People's Science Movement

Contents

Introduction	11
SECTION I **Knowledge for All:** **Capital vs. the People**	
1. The Production of Knowledge in Neoliberal Times	19
2. The Knowledge of Science and Technology as Commons	31
3. The COVID Pandemic Experience: Who Won, Who Lost?	57
SECTION II **Paradigm Shifts in Technology**	
4. Understanding the Philosophy of Technology	87
5. Restoring Conceptual Independence to Technology	95
6. The Dynamics of Technology and Self-Reliance	114
SECTION III **Mapping Public-Interest Science and Technology**	
7. Science in the Light of Social History	151
8. The Untold Story of the Left in Indian Science	183
9. Technology in a Postcolonial Setting: Notes from the Subcontinent	202
SECTION IV **Planning the Republic of Reason**	
10. Hindutva, Pythagoras and the Zero	211
11. Building a Nation with a Scientific Vision	224
Acknowledgements	256

'. . . freedom is the recognition of necessity;
science is the cognition of necessity'

D.D. KOSAMBI

Exasperating Essays

Introduction

How do we look at science and technology? What role do they play in society and, equally important, what is society's role in developing science and technology? These are questions that have engaged me as a social/political activist; but also, and perhaps primarily, because of my love for science and technology. It is this love that led me to be part of the people's science movement and the free software movement. This set of essays, several of them written over the decades, in earlier versions with different purposes, brings together all three interests—science, technology, and the society in which they are located. The earlier essays have, of course, been substantially revised or rewritten for the present times, keeping in mind new technologies and approaches, improved understanding, as well as fresh problems.

There is an older problem I must mention here. Unlike scientists, very few engineers or technologists write about the nature of their discipline. The scientists of an earlier era had an advantage: they came from a university background and were also taught philosophy. At least Einstein's generation of scientists were. Engineers, particularly in the British tradition, came from what we might essentially consider trade schools. In Victorian England, since engineers worked with their hands, they were not 'gentlemen'; they belonged to the lower orders and so very few practitioners wrote about the nature of their discipline. In 1959, when C.P. Snow spoke of the *Two Cultures,* the two different worlds of humanities and the sciences, technology remained excluded from these worlds and was not considered part of culture. The only exception here were the anthropologists, who had a broader definition of culture than the narrow gentlemanly one, as I discovered researching in JNU/INSDOC libraries, particularly the journal *Technology and Culture*.

In this book, I explore the complex relationship between science and technology. It is a given that both fields interact and that both are determined by society, but my focus is on what technology is. A part of this exercise is reactive: I react as an engineer, against treating technology as a sideshow of science. This was a common view at one time and remains current with a number of philosophers and historians of science. Vincenti, a technologist turned philosopher, suggests that this is because most philosophers who analyse technology have a background in sciences. If they were technologists, would they not be arguing whether science is not, at least in part, theoretical technology?[1]

A short essay I wrote more than four decades ago reflects my position on the relationship between science and technology even today, though I may not state it in the same words. The essay appeared in a journal called *Water World*, published from Delhi, and is included in this volume for its historical interest—how the philosophy of technology was viewed then, and how the trend continues even today.[2]

The objective of science is to know nature. Technology, however, starts with the objective of building an artefact; in other words, of changing nature. To build something in the real world, technologists need to bridge the gap between what is known and what is not. This is empirical knowledge that is

---

[1] Walter G. Vincenti, 'Engineering Knowledge, Type of Design, and Level of Hierarchy: Further Thoughts About *What Engineers Know*....' In Peter Kroes, Martijn Bakker (eds.) 'Technological Development and Science in the Industrial Age'. *Boston Studies in the Philosophy of Science*, Vol. 144. Springer, Dordrecht, 1992.
https://doi.org/10.1007/978-94-015-8010-6_2

[2] Prabir Purkayastha, 'Towards an Understanding of the Philosophy of Technology', *Water World*, Vol. 1, 1, pp. 20–23, July-September 1978.

codified in charts and tables, and comes from their practice.

Finally, in its pursuit of functionality—building artefacts with a prescribed and limited use—technology can sometimes also restate the fundamental relationships of nature, restatements that proceed into and impact science as well. For instance, originally a telegraph operator, Oliver Heaviside is the archetypical technologist: while solving problems related to telegraphy, Heaviside *changed* science by reformulating Maxwell's laws of electricity in the form we know today.[3] He also *changed* mathematics by solving differential equations as algebraic equations.

Increasingly, science needs artefacts in order to understand nature. The Hadron Collider is a perfect example, a $4.75 billion technological artefact—today's equivalent of the telescope and microscope which revolutionised astronomy and the life sciences. The objective of this artefact called the Hadron Collider is to perform experiments and advance our knowledge of nature. Equally, technology depends on science. Building the latest $300 million ASML equipment, required for the next generation of chip manufacture, needs a vast amount of scientific knowledge to be incorporated in the equipment. The knowledge incorporated in the ASML EUV lithographic machine is a precondition for producing chips that pack more processing power.

Why should we not combine science and technology under a common name? Why bother with separate labels? The answer is that their objectives are different. Just as Galileo's astronomy was not the product of the telescope; nor James Watt's steam engine the product of the knowledge of thermodynamics.

---

[3] Bruce J. Hunt, 'Oliver Heaviside: A First-rate Oddity', *Physics Today*. Vol. 65, 11, p. 48. He remains a relatively unknown figure even today in spite of his major work in physics and mathematics.
https://physicstoday.scitation.org/doi/10.1063/PT.3.1788

It is not as if this failure to recognise and locate the world of technology is the failing of a few societies. It goes deeper, to a split of perception and value between the head and the hand. From the slave societies of Greece and Rome, to the caste-driven society of India, the world of artefacts has been separated from the world of ideas. Scientists and mathematicians could enter the world of ideas, but not those who worked with their hands. The European aristocracy, first slave-owning and later land-owning, despised labour and, along with it, the instrument of labour—technology. So did the upholders of the caste system.

Having looked at different aspects of knowledge, both as science and as technology, the other set of questions I have addressed is that of the ownership of knowledge. If it is the common heritage of humankind, then science and technology as knowledge should belong to all of us, as human knowledge in the global commons. And if so, how should we regard patents or other forms of property rights over knowledge? Are they forms of enclosure of the commons? A repeat of how the commons were enclosed by the British aristocracy in the eighteenth century? The absurdity of the patents regime, particularly in medicine, is that the people are made to pay twice over for Big Pharma's patents; once by paying for the research conducted almost entirely in public institutions, and, again, by buying high-cost patented medicine.

The history of patents in India is illuminating in this context. An important part of the story is how we freed ourselves from the exorbitant price of medicines by using the Patent Act, 1970, and the Council of Scientific and Industrial Research (CSIR) laboratories. The AIDS epidemic taught us that new diseases remain a global threat, and that high-cost, patented medicine is a denial of the right to life for large sections of the people. The recent COVID-19 epidemic, followed by monkeypox, is teaching us that enclosing knowledge is not simply about Big Pharma and its profits, but a life-and-death question for people; as are patents for a whole slew of cancer and other drugs.

In this set of essays, I have tried to bring out the relationship among the triad of science, technology and society, and how their inner dynamics and their interrelationships drive their development, or the lack of it. I have gone back to the independence movement to see how this relationship has developed historically in India, and how ideology and political power has reshaped not only society, but also science and technology. Again, the historical accounts of technological change—in this case from self-reliance to a model based on multi-national corporations—have been mostly framed as economics. How do we assess change from within the frame of technology, not simply as an economic policy but as a technology policy?

When we talk of science, we refer to the effort to understand how nature works. Science, we can argue, is objective, dispassionate and rational. But when it comes to technology, even with the argument that it is for the benefit of the people, we encounter a problem: 'people' are not homogeneous. With any change, some people may lose out and some people may gain. We may believe that, in the long run, society as a whole will gain from change, but that is no consolation to the person who loses his or her livelihood in the process. While making a technological choice, we do privilege some sections and disadvantage others. In other words, technology choices are also social choices. This is why the choice of technology cannot be left to a technocratic elite.

It is a societal choice. It is also why we need a people's science movement, demystifying science and technology choices. Such movements help us perceive and weigh the social and political implications of the choices we make—whether these choices concern nuclear plants, or genetically modified food, or global warming.

These essays are not meant to be authoritative texts on the subject. They are, often, my idiosyncratic opinions on many issues. Sometimes, the views are articulated as a part of a public struggle. Sometimes, they are simply the opinions of a practising engineer, trying to explain, possibly to myself, what it is people like me do for a living.

Section I

Knowledge for All:

Capital vs. the People

1. The Production of Knowledge in Neoliberal Times

Science and nature have always had a vital link, in that science is viewed as the knowledge of nature. This knowledge gleaned from nature helps us develop our widely-used artefacts, such as tools, machines, or medicines. Scientists claim their discoveries have been guided by the thirst for knowledge, implying that their research aims at uncovering the secrets of nature. While this may have been the stated goal, science was never simply a disinterested quest for pure knowledge. Social needs, quite often the needs of the ruling classes—and under capitalism, the market—have been major drivers of the institutions of science. Nor do the laws of nature convert readily into artefacts; technological knowledge is also necessary to produce them.

This state of affairs led to the categorisation of science into two kinds: pure science and applied science, with science creating the knowledge that becomes 'applied science' when applied to our social needs. In such a simple and reductive scenario, technology becomes a mere application of science to social problems, while science loses sight of technology as an independent human activity.[1] In any case, scientific activity does not remain 'pure' or unsullied by commerce for long. Even number theory, once thought to be far removed from commerce, is now central to cryptography and part of the militarisation of knowledge. In his essay 'A Mathematician's Apology', G.H. Hardy wrote, 'Real mathematics has no effects on war. No one has yet discovered any warlike purpose to be served by the theory of numbers or relativity, and it seems very unlikely that anyone will do so for many years.'[2] He was proved wrong

---

[1] I have dealt with the relationship between science and technology in Section II, Chapters 4 and 5.
[2] G.H. Hardy, 'A Mathematician's Apology', 1940.

quite soon. During the Second World War, British mathematician Alan Turing broke the code of the 'Enigma'—cipher codes used by Germany and its allies—incidentally illustrating the 'value' of the theory of numbers to the military. And relativity plays a key role in GPS (global positioning system) devices, of which the cell phone is the most popular. The university system that developed in Europe looked upon science as a key element in the drive to industrialise nations. 'By the beginning of the nineteenth century,' write Dhruv Raina and Prakriti Bhargava, 'two large structures for higher education emerged in revolutionary Europe . . . These were the German university of teaching and research and the French model of the *grandes écoles* or the specialist technical institutions.'[3] The German Humboldtian model of the university had an impact on the English and American universities as well. Both the French and German universities had close ties with industries in their countries. This phase of developing knowledge within the structure of educational institutions marked a break from the earlier form, that of scientific societies, which, quite often under royal patronage, had provided a space for scientists to come together and exchange knowledge. The universities created a space where science and knowledge development could become an occupation, rather than an aristocratic activity or one dependent on the patronage of rich aristocrats.

In Great Britain, the universities—Cambridge and Oxford—had begun with teaching religion to the younger sons of elite landed families, preparing them for duties in the church; and later on so they could serve in the colonial administration.[4] Despite the insistence of the landed elite that British universities remain

      https://archive.org/details/hardy_annotated
[3] Dhruv Raina and Prakriti Bhargava, 'Rethinking the University of Teaching and Research', in C. Shambu Prasad, John D' Souza (eds.) *Rethinking Universities for Development Intermediaries Innovation & Inclusion*, pp. 26–37, 2013.
      http://ced.org.in/docs/kics/UNIID/Rethinking_Universities_papers.html
[4] Stefan Collini, *Speaking of Universities*, Verso Books, p. 37, 2017.

focussed on the 'classics', science and technology eventually grew in importance even there.[5] Universities in India were supposed to produce the clerical staff required to manage the day-to-day activities of the empire. Unfortunately for the British, the universities also laid the seeds of Indian nationalism. The British did not make this mistake again; they set up universities in Africa only in the 1940s.[6] Before that, the African elite were encouraged to enrol in British universities so they might return with an exalted image of European, particularly English, civilisation. The French had the colonial elite trained in Paris for similar reasons. In the twentieth century, state support for higher education and research emerged as the main source of funds, though the university and research institutions still had some autonomy relative to the state.[7] Viewing the university as a part of civil society, and examining the role it plays in the larger struggle for hegemony over society, comprise a major subject. For our purpose, we should see universities as a seat of class struggle, an arena for the battle of ideas. The relative autonomy of the university and the space to develop critical views on various subjects makes it possible for the

---

[5] Charles P. Snow, *The Two Cultures and the Scientific Revolution*, Cambridge University Press, 1959. In his influential work Snow lamented the 'great cultural divide' between science and the arts, and condemned the British educational system for emphasising the Greek and Latin classics at the expense of science and technology. He pointed out that this deprived the British elite of the education necessary for them to manage the modern scientific world.

[6] Appollos O. Nwauwa, 'The British Establishment of Universities in Tropical Africa, 1920–1948: A Reaction Against the Spread of American "Radical" Influence', *Cahiers d'études africaines*, pp. 247–274, 1993.
https://www.persee.fr/doc/cea_0008-0055_1993_num_33_130_1520

[7] Though the US had many private universities well-endowed with funds, in science and technology, government grants, particularly for the military, continue to be a major source for scientific institutions. See, for example, Robert D. Atkinson and Kevin Gawora, 'U.S. University R&D Funding Falls Further Behind OECD Peers', *Information Technology & Innovation Foundation*, 2021.
https://itif.org/publications/2021/04/12/us-university-rd-funding-falls-further-behind-oecd-peers/

university to generate new ideas as well as practical advances in the sciences. Without this relative autonomy of institutions and the nurturing of a critical outlook, it would have been difficult for the basic sciences to advance. And without advances in the basic sciences, the application of new science to technology would not develop. The widely understood need was for the state to be involved in basic research, while industry could develop the applications of this research more or less on its own. This still forms the underlying reason why the state funds basic research, and why science and applied sciences need state support for educational and research institutions.

This neat division of basic sciences and applied sciences, if it ever really obtained, has disappeared today. Yet, the division continues to underpin a lot of writing on science and technology, where technology is viewed as applied science.[8] Even within this framework, many advances that would have been called basic sciences earlier—for example, in particle physics and biology—are converted into marketable applications very quickly. The reverse is equally true: for any advance in particle physics, we have to invest a huge amount of capital in building new instruments of enquiry— the Hadron Collider or the Webb Telescope. Confirmation of the existence of the Higgs boson took 48 years, and only happened after $8 billion were invested to create the Large Hadron Collider at CERN.[9] While this advances our knowledge of nature, it may well have no immediate commercial utility. The other aspect of Big Science, of which the Large Hadron Collider is one example, is the industrial scale of any experiment that uses it. The two Higgs boson teams in CERN, which operated the two detectors of the

---

[8] I have covered this partly in Section III, and more fully in Section II.
[9] The mechanism that gives rise to the Higgs boson was postulated in three papers in 1964, published in *Physical Review Letters*. Two of the six authors, Peter Higgs and François Englert, got the Nobel Prize in Physics in 2013. On 4 July 2012, the ATLAS and CMS experiments at CERN's Large Hadron Collider announced they had each observed a new particle in the mass region around 125 GeV, or the long-predicted Higgs boson.

Collider, wrote a joint paper in 2015 which had 5,154 authors from 45 countries and 245 institutions from all over the world, including India.[10] This is science being done at an industrial scale. There are two transformations taking place today in educational institutions. One is to treat educational activity like any other business. The quantification of output in terms of teaching hours, students' performance, research papers, contributions to the income of the university from business and students, or 'key performance indicators', have become the yardstick for judging teachers, departments, and institutions. In this view, an educational institution should be treated like any other commercial activity. Its objective is to maximise output—in terms of more students, more patents, consultancies, research grants—while minimising costs.

The second transformation is the privatisation of the output of institutional research, even when publicly funded or funded by government grants. This generates more income for the university, while a still larger share of benefits flows to the corporations who are able to 'enclose' the output by buying patents on the findings and applications of such publicly funded research. Such instances are most visible in the case of the pharmaceutical industry. They are facilitated by the state's decision that the output of scientific or technological research, even if supported by the state, should be handed over to private capital.

Increasingly, markets and the demands of global capital control science and its advances. Neither the objectives of advancing the knowledge system, nor those of meeting the needs of the people, are served by the current practice of science. Knowledge, and larger social goals, are sacrificed to the neoliberal economic order which values immediate gain as the driver of science. Science as an open system, at least among scientists, is giving way to the logic of capitalist enterprise. While we may talk about the knowledge

---

[10] Davide Castelvecchi, 'Physics Paper Sets Record with More Than 5,000 Authors', *Nature*, 15 May 2015. https://doi.org/10.1038/nature.2015.17567

economy, the output of scientific research has been privatised, even when publicly funded within the heart of the educational system. Before a result is published, scientists are already busy with establishing the claim to prior knowledge indispensable for filing patents. The output of scientific research no longer comprises research papers in journals that increase the boundaries of knowledge, but patents which can be turned into money.

This has led to a new goal within the scientific community: creating a new class of scientists who behave as entrepreneurs. They often operate within the public science institutions, but have become science entrepreneurs using the output of their research. The goal of research is no longer the production of knowledge but the creation of monopolies for either private capital or for the 'scientist as entrepreneur'.

Monopoly over knowledge, whether used to sell a software program, a medicine, or a seed, translates into the ability to extract super profits. It also aborts the possibility of science being developed through an open network. It is paradoxical that while the internet and open access to knowledge have made it possible for large groups of people around the world to work together and bring about major advances in science, we are also witnessing a time in which knowledge is packed into cubbyholes, precluding cooperation.

The free software movement of the 1980s showed the power of new networked structures in creating software, and their superiority over privatised software development. Never before has society had the ability to bring together different communities and resources on this scale. We have seen what collaboration is capable of when harnessed to scientific activities such as particle physics or astronomy.[11] So, what stands in the way of liberating this enormous power of working together for the generation of new knowledge and new artefacts? Clearly, the impediments are

[11] Take Higgs boson or the hunt for black holes, these activities have seen some of the best open and collaborative science in recent times.

the monopoly rights and private appropriation inherent in the intellectual property rights (IPR) order.[12] The understanding that science needs to be restored as an open and collaborative exercise has given birth to the commons movement. While the environmental and ecological movements have also fought against the privatisation of the commons, the kind of commons they looked at are finite resources, such as grazing land, forests, fisheries, oceans, and the atmosphere. These commons are natural resources, once thought to be infinite, and now understood to be finite, capable of over-exploitation and degradation. The knowledge commons are intrinsically different in that they—a law of nature, let's say, or knowledge of a genetic code—do not degrade with repeated use. Open use on a large scale enriches the knowledge commons.

In this 'commons' view of the world, intellectual property rights are attempts to exclude people from the domain of knowledge by enclosing it, in a manner similar to the enclosure of the village commons carried out over the last 500 years. It uses a legal artifice called IPR to privatise knowledge that was always publicly held. Any enclosure of knowledge is doubly pernicious: it not only restricts access to others, but also puts a price on access to something which is infinitely duplicable and does not degrade with use. The enclosure of knowledge using the IPR regime is even more iniquitous than the earlier kind of enclosure movements, which had argued that private property rights prevented the degradation of the commons. The struggle against intellectual property rights of various kinds becomes a battle to preserve the global commons, specifically knowledge in its various forms.

The earlier system of developing scientific knowledge resided primarily within the structures of higher education. Universities,

---

[12] Arti K. Rai, Jerome H. Reichman, Paul F. Uhlir, Colin R. Crossman, 'Pathways Across the Valley of Death: Novel Intellectual Property Strategies for Accelerated Drug Discovery', *Yale Journal of Health Policy, Law and Ethics*, Vol. 8, 1, pp. 53–89, 2008. https://ssrn.com/abstract=1085027

colleges, and other institutions of higher learning were the centres where new advances in science originated. As these centres of education were relatively autonomous of both state and market, the system of generating new knowledge was not constrained by the immediate class needs of society. This produced within the university system a sense of independence and self-regulation, and the education given to students had a purpose larger than merely serving capital or the needs of the state. This is why the educational system became, additionally, a space where new ideas arose not only in the various disciplines, but also about society itself.

The humanist view of science and technology fitted very well within this overall structure. Science was supposed to produce new knowledge, which could then be mined by technology to produce artefacts. The role of innovation was to convert ideas into artefacts—thence the patenting system that provides protection to useful ideas embodied in the artefacts. Artefacts could be patented, but not knowledge of nature.

The transformation of this system after more than a hundred years has come from two distinct sources. One is that science and technology are far more closely integrated than before, and scientific advances can be turned rapidly into marketable products. For example, advances in biology that create new pharmaceutical compounds,[13] or advances in physics entering the latest chip-making machines,[14] happen much faster today than they did earlier.

---

[13] The mRNA COVID-19 vaccines exemplify public domain science being converted to private property. Massimo Florio, 'Patents for COVID-19 Vaccines are Based on Public Research: A Case Study on the Privatization of Knowledge', *Working paper CIRIEC No. 2021/03*, 2021. https://www.ciriec.uliege.be/repec/WP21-03.pdf

[14] The key to the production of chips under 10 nm is the ability to use extreme ultraviolet lithographic machines. The manufacturer is a Dutch company, ASML. The combination of advanced physics in many aspects of the machines made this sub10 nm technology possible. Advanced scientific research, along with the capability to integrate them in the EUV lithography machine, are not contributions made only by ASML. It is the result of its cooperation with 180 universities in Europe and the US. Kailey Erlich, 'Research University Partnerships Support Future of Technology', *ASML*, 7

Unfortunately, market fundamentalists the world over are pushing measures similar to the Bayh-Dole (Patent and Trademark Law Amendments) Act,[15] to enclose the advances in knowledge created by public funding. The Bayh-Dole Act, introduced in the United States in 1980, reversed the almost universal assumption that public-funded research should not be protected by private rights in the form of intellectual property protection. The Act allowed universities and other non-profit entities to patent research funded from public resources. It created the conditions for the university system in the United States to work much more closely with large corporations. Publicly funded research can now be privatised by the university, by selling the patents to private corporations. The people pay twice: once for funding research with public money, and the second time by paying a high price for products developed from publicly funded research.

Other factors have also contributed to transforming the system of knowledge production, most notably the neoliberal policies adopted across the globe. Public funding on research has suffered, especially in developing nations, as neoliberal economics has led to a general squeeze on government finances. This, coupled with a reliance on the market, has given rise to the notion that the way forward is to source research funding from the private sector. There is evidence today that private funding of research distorts the priorities upheld by public funding. Private companies are able to exercise decisive control over the trajectory of research by funding only select parts of a research project. And, of course, the fruits of public-funded research are placed at their disposal.

The weakening, if not the demise, of the UN system has also shifted emphasis away from public-funded science that embodied the notion of collaboration amongst nation states. Today, the UN

April 2021. https://www.asml.com/en/news/stories/2021/research-partnerships-with-universities

[15] Clifton Leaf, 'The Law of Unintended Consequences', *Fortune*, 19 September 2005.

system promotes 'public-private partnerships'. For example, the major share of the World Health Organisation's budget comes from private foundations and donors, or from governments in the global North who wish to fund specific programs that are of interest to them.

The enclosure of the commons is taking place not just in the sciences, but also in traditional knowledge. Pharmaceutical companies, for instance, appropriate community-based knowledge, privatising it in various forms. This pertains both to the specific knowledge and practices of communities as well as the biological resources they nurture. The struggle to protect the rights of such communities is also a struggle to protect traditional knowledge as commons. These commons are not public domain but the common property of a group, and hence allow for community rights as opposed to the property rights of individuals or corporations. The commons license approach is increasingly considered a means of protecting traditional knowledge.

While the growing disconnect between the production of knowledge and its private appropriation needs to be scrutinised, new ways of knowledge production point to various possibilities for democratising the practice of knowledge. The information technology sector has shown today that new technologies and methodologies can be developed by cooperative communities. It may be argued that this sector is unique in that the 'reproduction costs' of its 'artefacts'—the software—are relatively low. The question is whether such approaches can be designed for other areas, such as the life sciences. Is it possible to innovate ways of establishing 'creative commons' in which new technologies and methodologies are developed by cooperative communities? The issue here is not simply of keeping scientific knowledge as open access, or some kind of science-as-commons, or open science. It is also how to *develop* science as commons through the common activity of a group. It takes for granted that the output of such groups will be open access or a science common. It concerns not simply the reproduction rights to knowledge but also how to produce science.

The second question we need to address is how to bring back societal concerns into science institutions. How do we democratise these institutions so that larger social goals determine the priorities of science? Would research into diseases that affect the poor be funded by a corporate sector that is uninterested in developing medicines for people who cannot pay? How do we take on board the concerns of poorer countries who have neither the money nor the scientific resources to address such problems? How do we introduce and promote equity in the system of advancing scientific knowledge?

This leads us directly to the larger issue of how society, as a whole, exercises control over the enterprise of science. If science today is a major economic force, the larger goal of democracy and equity in society will inevitably include science. It is not surprising that a number of crucial questions in today's world require an understanding of science. In the absence of such an understanding, a few scientists from the ruling establishment would get to impose their judgements as the 'scientific' decisions for society.

Earlier movements of scientists had placed this issue within the context of the social responsibility of the scientist. The scientist, in this view, owes it to society to be conscious of his or her activities and bring them to public notice. The scientist had a two-fold responsibility—understanding the implications of science for society, and also becoming an active champion for the right kind of science and application. Race science, as in debates about IQ, is a stark reminder of the need to determine 'what kind of science'. Whether a nuclear bomb should exist or not is about examining 'how science should be applied'.[16] The role of scientists in nuclear

---

[16] Seventy of the scientists working in the Manhattan Project signed the appeal to President Truman not to use the nuclear bomb on Japan. This effort was spearheaded by Leo Szilard and Einstein. Subsequently, Einstein and Bertrand Russell signed the manifesto in July 1955 to warn the world against nuclear war. Pete, 'The Manhattan Project Radicals and the Birth of

disarmament is perhaps the most important example of this earlier work. The scientific workers movement—the movement for popularising science that developed in the 1940s and 1950s—grew out of this perspective. As did the people's science movement in India, and the science for the people movement in the United States in the 1960s and 1970s: both still continue.[17] Today, the need for organising scientists to struggle for a more democratic scientific decision-making process must go hand-in-hand with a strong movement to bring science to the people. If global warming is to be combatted, or nuclear disarmament pursued, it is not enough for the scientists to say so. Science has to be brought out of the ivory tower and de-mystified so that the people affected by such decisions can also assert their voice. Science is too serious a business to be left to the scientists—it must be a part of our larger struggle for equity and democracy in society. We need to fight, so the products of science and technology serve the common interest and are accessible to all, and we need to involve the people in this struggle. This is the challenge before us today.

---

the Anti-Nuclear Movement', *Radical Tea Towel*, 6 August 2019. https://radicalteatowel.co.uk/blog/the-manhattan-project-radicals-and-the-birth-of-the-antinuclear-movement

[17] See Chapter 8 (Section III) for more on this history.

2. The Knowledge of Science and Technology as Commons

They hang the man and flog the woman
That steal the goose from off the common,
But let the greater villain loose
That steals the common from the goose.

English folk poem, circa 1764[1]

One of the key determinants of today's world is the speed with which innovation takes place and is brought within the sphere of production.[2] The growth of technology is a continuous driver of the economy. There has been considerable discussion on the monopoly created through the 'reproduction' of an innovation via patents; but relatively less attention has been paid to the way innovation takes place, and the overarching structures that facilitate or retard its growth. Does today's networked world hold new possibilities for creating knowledge and innovation? Are these possibilities currently impeded by the patents model of incentivising innovation? If so, can we expand the notion of the 'commons' to develop such possibilities?

The model of generating innovations in technology was conceived to be 'private' from the start. This is unlike the case of science—public science has been the major driver of science research. The function of public science referred to here is that of

---

[1] David Bollier, *Silent Theft: The Private Plunder of Our Common Wealth*, Routledge, 2003.
[2] See discussions on technology and the pace of innovation in Prabir Purkayastha, 'Technology: Breaking the Cycle' in *First Academic Seminar of the IBSA Dialogue Forum*, Brasilia, 12 September 2006.
http://livros01.livrosgratis.com.br/al000087.pdf

advancing knowledge, not disseminating knowledge, which is its other function. The patenting system originated in the days of the lone inventor and the need for society to prevent an invention from dying with the inventor.[3] In exchange for disclosing the details of the invention, the inventor was granted a temporary monopoly, or a patent. The inventor had to make knowledge of the invention public by submitting a detailed description to the patent office. The description would enable reproduction of the invention by others once the patent period, then of fourteen years, was over. It appears that Richard Arkwright had his patent on the spinning jenny cancelled because his description was incomplete. Patents were—and still are—a source of knowledge for the technical community.

The need for patents has always been viewed as a necessary evil. The US Constitution allows the Congress 'to promote the progress of science and useful arts, by securing for limited times to authors and inventors the exclusive right to their respective writings and discoveries'.[4] Thus, even in the US, this exclusive or monopoly right is not granted because the inventor somehow owns the idea embodied in the patent, but so as to promote science and technology—in other words, larger societal goals. Historically, the lone inventor gave way in the early twentieth century to large corporate or state-funded research laboratories. Big capital has increasingly looked to public science institutions and universities rather than industrial laboratories to produce the knowledge they require. With the Bayh-Dole (Patent and Trademark Law

---

[3] In Britain, patents originate from any monopoly that the crown gave for trade and manufacture, including inventions, using letters patent or an open letter. The knowledge of the invention had to be transferred through apprentices or, later, through a written description. Peter Wells and Tilaye Terrefe, 'A Brief History of the Evolution of the Patent of Invention in England', *Intellectual Property Institute of Canada*, Vol. 35, pp. 66–77, 2020. https://mcmillan.ca/wp-content/uploads/2021/11/A-Brief-History-of-the-Evolution-of-the-patent-of-Invention-in-England.pdf

[4] Article I, Section 8, Clause 8 of the United States Constitution. https://www.law.cornell.edu/constitution/articlei#section8 https://www.law.cornell.edu/wex/intellectual_property_clause

The Knowledge of Science and Technology as Commons 33

Amendments) legislation of 1980, which essentially allowed private capital to capture the output of publicly funded research, this model has come to dominate public-funded science in the US.[5] In India, the example of the US has gained ground as worthy of emulation, and other countries have followed suit to varying degrees.Interestingly, this is also the time in which alternate models of generating knowledge and innovation have gained ground.[6]

The free software movement, launched in the mid-1980s, has shown that networked and open collaborations among 'hackers' can produce software of far better quality than well-heeled corporations working in isolation. Hackers are those in the software community who can hack *code*; they do not break into computers illegally as often portrayed by the journalistic interpretation of hacking.[7] Visible in this model is the power of open, collaborative structures, working without so-called material incentives. The free software movement has resurrected older models that have played key roles in innovation and technology development, as with the steam engine in Cornish mines and the blast furnace in Great Britain and the US.[8] Alessandro Nuvolari argues that 'together with individual inventors, collective invention settings were a crucial

---

[5] An accessible critique of the Bayh-Dole Act is by Clifton Leaf, 'The Law of Unintended Consequences', *Fortune*, 19 September 2005.
https://money.cnn.com/magazines/fortune/fortune_archive/2005/09/19/8272884/index.htm
A more detailed analysis can be found in David C. Mowery, Richard R. Nelson, Bhaven N. Sampat, and Arvids A. Ziedonis, *Ivory Tower and Industrial Innovation: University-Industry Technology Transfer Before and After the Bayh-Dole Act*, Stanford University Press, 2004.

[6] Prabir Purkayastha, Satyajit Rath, Amit Sengupta, 'Looking at Knowledge and Science as Commons', Background Paper, Workshop on Science Commons, Delhi, 18 January 2008.

[7] Vivek Shangari, 'Hackers and the Open Source Revolution', *Open Source For You*, 25 February 2012.
https://www.opensourceforu.com/2012/02/not-crackers-but-hackers-open-source-revolution/

[8] Robert C. Allen, 'Collective Invention', *Journal of Economic Behavior & Organization*, Vol. 4, 1, pp. 1-24, 1983.
https://doi.org/10.1016/0167-2681(83)90023-9

source of innovation during the early phases of industrialisation'.[9]

The case of the Cornish mines in the development of steam engines is particularly interesting, as the Cornish miners regarded the patent of Matthew Boulton and James Watt a serious impediment to technological progress, and, as a consequence, they did not patent their own advances. The question to explore is that if we accept the concept of a knowledge economy, what instruments are most appropriate for the expansion of this economy, especially for developing countries such as India? Is there evidence to believe that incentives for innovation require a strong patenting regime? Or is this one of those claims that turn into 'truth' through repetition?

REPRODUCTION OF INNOVATION: PATENTS, COPYRIGHTS, AND WTO/TRIPS

The last few decades have seen the creation of a new category of private property rights called intellectual property rights, bringing under one umbrella what were earlier disparate property rights.[10] Thus different kinds of private property rights—the creative rights of authors under copyright, and industrial property rights such as patents, trademarks, trade secrets and industrial designs—have been brought under the common rubric of intellectual property rights (IPR) in the new WTO regime as Trade Related Intellectual Property Rights or the TRIPS Agreement.[11] The objective of this renaming exercise was two-fold. First, it sought to legitimise what are essentially corporate rights by bringing them under the cover

---

[9] Alessandro Nuvolari, 'Collective Invention During the British Industrial Revolution: The Case of the Cornish Pumping Engine', Eindhoven Centre for Innovation Studies, *ECIS Working Paper Series*; Vol. 200402. Technische Universiteit Eindhoven, 2004.
https://pure.tue.nl/ws/files/2343508/669221.pdf

[10] S.P. Shukla, 'From GATT and WTO to the Patents Act, 2005: The Long Arc of Resistance', in *Political Journeys in Health*, LeftWord Books, 2021.

[11] Ibid.

of property rights. The right of individual creativity, for example, of an author in terms of her/his authorship, which is inalienable, is quite distinct from copyright which can be bought or sold. The second was to expand enormously the scope of such rights, primarily of Big Capital. The battle to expand the property rights of Big Capital has been a continuous one. A notorious example is that of the US copyright law being extended every time Disney's copyright of Mickey Mouse reached its expiry date.[12] This is why the Copyright Act of the US is derisively referred to as the Mickey Mouse Law. The Disney empire, worth about $200 billion today, was based on the huge success of the Mickey Mouse character. This is ironic, given that the first Mickey Mouse movie to be commercially released, *Steamboat Willie* (1928), not only parodied the title of the Buster Keaton hit *Steamboat Bill, Jr* (also 1928), but also used at least two older and popular songs.

The impact of this new IPR regime, coupled with the global trading regime under the World Trade Organisation, led to the private appropriation, on a grand scale, of biological and knowledge resources held in common by society. Today, this regime has expanded to the patenting of life forms, genetic resources, genetic information in life sciences, patenting methods and algorithms in computational sciences, and even to patenting how business is done. Not only are methods and algorithms patented, but copyright has also been extended to software and all forms of electronically held information. The traditional knowledge and biological resources held and nurtured by different communities are being pirated by global corporations. More and more, the enterprise of science as a collaborative and open activity to create open knowledge is being altered into a corporate exercise to create monopolies and milk super profits from the consumers.

---

[12] Lawrence Lessig, 'The Creative Commons', *Montana Law Review*, Vol. 65, 1, 2004.
 https://scholarworks.umt.edu/mlr/vol65/iss1/1

This happens even if the knowledge has been created by publicly funded research in the first place.

The impact of such appropriation is now visible. The HIV/AIDS epidemic has shown that what stands between life and death for the infected is the profit of Big Pharma.[13] The COVID-19 pandemic shows that AIDS was not an isolated instance, the crisis remains an ongoing one. It is impossible for the vast majority of people in the world today to pay the costs of new lifesaving drugs which are patent-protected. COVID-19 has also shown us that apart from drugs, we must look at vaccines and diagnostics, along with other forms of intellectual property rights such as trade secrets that are important for their production.If the IPR regime has been damaging the lives of those suffering from disease, what lies in store for agriculture is even worse. With biotechnology and bioinformatics, corporate seed companies and corporate plant breeders are poised to control global agriculture and food production. With food prices already skyrocketing, the impact of such a monopoly on vast sections of the people can well be imagined.

The battle over global commons has also been waged at the level of ideology or ideas. We have the famous 'Tragedy of the Commons', written by Garrett Hardin,[14] essentially an ideological attack on the commons. In Hardin's view, the commons suffer from the dual problem of over-consumption and underinvestment: every user of the commons has an incentive to consume more, and no incentive at all to invest in renewing the commons. His solution was what actually happened in England—the enclosure of the commons for private use. The other viewpoint, put forward by Elinor Ostrom in her various works[15] on the commons is that, in an overwhelming

[13] For more on this, see Chapters 3 and 8.
[14] Garrett Hardin, 'The Tragedy of the Commons', *American Association for the Advancement of Science*, Vol. 162, 3859, pp. 1243–1248, 1968. https://www.jstor.org/stable/1724745
[15] Ostrom's Nobel Prize citation for Economics in 2009 describes her work on small communities managing shared resources as disproving the view

number of such cases, the people using common resources create voluntary structures that regulate the commons, thus preventing Hardin's 'tragedy'. These debates pertain more to the physical (or natural) commons: forests, air, and water, for example. Some of these commons are treated as if they were inexhaustible—air, for instance; we know, of course, that this is not true, given carbon emissions and their impact on global warming. It is ironic that the infinite commons of knowledge are treated as if they were finite, while the finite commons of air and oceans are treated as if they were infinite. Interestingly, it is not the small users, the citizens of the world, who overuse the commons, as Hardin had predicted.

The culprit is Big Capital, whose thirst for profits overrides its concern for humanity. Consider the example of Exxon. As early as 1977, Exxon's research showed that the world had only 5–10 years before hard decisions would have to be taken to limit the rise of temperature below 2–3 degrees centigrade.[16] Exxon not only buried this result but became a leading climate-change denier.

Knowledge as commons is quite distinct from the physical commons. Unlike the natural commons, a physical resource which can be depleted by overuse—such as grazing or fishing—knowledge does not get depleted with use. Intellectual property—as a property right on knowledge—is, in this sense, different from the natural commons and their use. The formula of a chemical compound that an inventor has produced in a laboratory does not get reduced if another chemist uses it. This is in contrast to a physical object, say, a sandwich. Knowledge is what economists and legal experts define (in their usual obscure language) as non-

---

'unanimously held among economists that natural resources that were collectively used by their users would be over-exploited and destroyed in the long-term.'
https://www.nobelprize.org/prizes/economic-sciences/2009/ostrom/facts/

[16] Shannon Hall, 'Exxon Knew About Climate Change Almost 40 Years Ago', *Scientific American*, 26 October 2015.
https://www.scientificamerican.com/article/exxon-knew-about-climate-change-almost-40-years-ago/

rivalrous goods. Thus intellectual property rights, given by the state as a *monopoly* to the inventor (or the copyright owner), is, in effect, a tax on the consumer. The consumer pays for this right through the higher price of the product.

The monopoly profits the inventor earns must then be balanced against the tax that the consumers pay. This is what Michael Heller and Rebecca Eisenberg examined with their paper (published in *Science* in 1998) on anti-commons in biomedical research: the possibility that instead of overuse of the commons, the proliferation of patents is leading to their underuse, as people are unable to pay for the high cost of the product.[17]

When patent holders fence in knowledge, they prevent its use by others; and this in turn leads to the underuse of such knowledge. The problem is compounded by patent trolls, who buy patents on the cheap and weaponise them by suing companies for infringement, in the hopes that many companies would pay rather than fight expensive legal battles. The proponents of a strong IPR regime claim that even if patents have social costs, they are useful to promote the innovations required by society. But even if we focus narrowly on the patenting costs of companies against benefits in terms of revenue, the figures indicate that the bang is not worth the buck involved in patenting. Researchers James Bessen and Michael Meurer have analysed the revenues generated from patents against the cost of filing, maintaining, and defending patents in courts.[18] They conclude that except in the case of pharmaceuticals, patents generate far more litigation costs than revenue. The numbers are clear: in the United States, the domestic litigation costs—$16 billion in 1999 alone—were about twice the revenue earned from patents. Almost two-thirds of all revenue came from pharmaceuticals and chemicals. Worse, the more innovative the company, the greater

---

[17] Michael Heller and Rebecca S. Eisenberg, 'Can Patents Deter Innovation? The Anticommons in Biomedical Research', *Science*, Vol. 280, p. 698, 1998. https://scholarship.law.columbia.edu/faculty_scholarship/1158

[18] James Bessen and Michael J. Meurer, *Patent Failure: How Judges, Bureaucrats, and Lawyers Put Innovators at Risk*, Princeton University Press, 2008.

the likelihood of its being sued. Software and business method patents fared the worst, with costs far outstripping the benefits of patenting.

Even if we leave aside the broader question—of whether societies benefit from greater innovation—and examine the very narrow one of whether companies that are innovative benefit from patenting, the answer is that they do not. This conclusion by Bessen and Meurer is no different from what others have discovered in the past: if patents did not already exist, they would be a poor way of rewarding innovation.

Apart from Bessen and Meurer, research by Michele Boldrin and David Levine also shows that patents do not promote innovation in societies.[19] Most of the historical data from countries with different forms of patent protection—strong protection versus weak protection—do not show significantly different rates of innovation. A number of recent cases on patents in the United States Supreme Court and in the US Federal Court have shown that companies investing heavily in advanced technologies, such as information technology or genomics, are currently moving away from the patent model.[20] A major exception to this trend are the big pharmaceutical companies. Even major software

---

[19] Michele Boldrin and David K. Levine, *Against Intellectual Monopoly*, Cambridge University Press, 2008.

[20] In *KSR vs Teleflex,* a key case on patent protection, a number of hi-tech companies regarded as innovative opposed the easy grant of patents. The exceptions were the pharma companies. The judgement raised the bar on patents. *KSR International vs Teleflex*, US Supreme Court, 2007. This was followed by *Alice Corp. vs CLS Bank International* in 2014, which significantly restricted the scope of software patents. In the 2013 ruling in *Association for Molecular Pathology (AMP) vs. Myriad Genetics Inc.*, the US Supreme Court declared that human genes cannot be patented as they are a 'product of nature'. There are ongoing attempts in the US Congress to set the clock back, so the battle continues. Kelly Servick, 'Controversial U.S. Bill Would Lift Supreme Court Ban on Patenting Human Genes', *Science*, 4 June 2019. https://www.science.org/content/article/controversial-us-bill-would-lift-supreme-court-ban-patenting-human-genes

companies have begun producing open and/or free software products, while companies such as Google, Facebook (Meta), and Uber, are significant contributors to free and open source software.[21] These companies provide services of a different kind. Google and Facebook (Meta) sell the audience/user as a commodity to advertisers; they do not sell software as a product. Why do these companies put their developments under various forms of commons license? They benefit from the larger software community that uses their tools, thus creating an eco-system that improves these tools. Such tools help their business, which is not selling software, but a different set of services. Google uses its search engine and YouTube, or Facebook (Meta) uses Facebook-Insta, to sell us advertisements. This is their primary revenue model. A product company like Microsoft sells software as a product; Apple sells iPhones in which its software is embedded as the product, and it considers software as the product or a part of its products. This is why not all big digital monopolies have a common approach to software and the commons. Public activism in software—the Free and Open Source Software (FOSS) Movement—has forcefully counterposed the concept of 'commons' to that of intellectual property rights. The FOSS movement's take on intellectual property rights is that IPR is an attempt to exclude people from the domain of knowledge by enclosing it. Through a legal artifice, knowledge is taken from the public domain and privatised. The struggle against IPR of various kinds then transforms into a battle to preserve the global commons, specifically knowledge in its various forms. Different forms of commons are used in different spheres, from software to drug discovery and agriculture.

[21] There are significant differences between how open source and free software communities look at how software should be licensed, but in the fight against privately owned software, either licensed or patented, they have a certain common understanding. 'Difference between Free Software and Open Source Software', *GeeksforGeeks*.
https://www.geeksforgeeks.org/difference-between-free-software-and-open-source-software/

The Knowledge of Science and Technology as Commons

A HISTORICAL LOOK AT PATENTS: CORNISH MINES

As an incentive, the patent gives a monopoly to the inventor for a certain period, in lieu of which s/he makes the invention public. In economic terms, this monopoly allows the patent holder to extract rent from all users: it is the state allowing the patent holder the right to levy a private tax. Naturally, the question arises whether patents (or monopolies) are the best form of providing such incentives.

Even if we accept that inventors need material incentives, incentives other than patent monopolies are available. These could include a royalty for the inventor from any producer who wanted to work the patent, but not a monopoly over all reproduction of the invention. One such option is the License of Rights where the patent holder can declare in the patent that it is available for licensing for a certain fee. Or the state could offer prizes from its kitty for socially useful inventions—this is a policy that a number of states have used to encourage inventors.

The question is whether the monopoly patent regime has indeed helped promote innovation. Suppose we start with the most celebrated innovation, the one described by textbooks as a critical factor in the industrial revolution: the steam engine. James Watt perfected his version of the steam engine and secured a patent in 1769. In 1775, using the influence of his rich and influential business partner, Matthew Boulton, Watt succeeded in getting parliament to pass an Act extending his patent till 1800. If we examine the developments in steam engines, we can decide whether the Watts patent helped promote innovation or actually stifled development.

The major beneficiary of advances in steam engines would have been the mining industry in Cornwall. Watt spent his entire time suing the Cornish miners if they tried to make any improvements to his design. The firm of Boulton and Watts did not even manufacture steam engines then; they only allowed others to construct the engines based on Watt's designs, for which they claimed huge royalties. If we examine the increased efficiencies of steam engines and plot them against the passage of time, we find that *after the initial Watts breakthrough*, during the period that Watt had monopoly, *all further improvements virtually stopped, starting again only after the expiry of his patents.*

During the period of Watt's patents, the UK added about 750 horsepower of steam engines per year. In the thirty years following the expiry of Watt's patents, additional horsepower was added at a rate of more than 4,000 per year. Moreover, the fuel efficiency of steam engines changed little during the period of Watt's patent; while between 1810 and 1835 it is estimated to have increased by a factor of five.[22] The major advance in steam engine efficiency took place not because of Watt's invention but in spite of him. Interestingly, all those who made further advances, such as Richard Trevithick (the inventor of the first steam locomotive), did not file patents. Instead, they worked on a collaborative model in which all advances were published in a journal called the *Lean's Engine Reporter*, collectively maintained by the mine engineers. This journal published best practices as well as all advances that were being made. This was the period that saw the fastest growth of engine efficiency.

Research reveals that there is little concrete evidence that increased patent protection actually helps innovation. In fact, in addition to the evidence of the Cornish mines, blast furnaces in nineteenth-century UK and the US show that collective innovation settings led to faster diffusion of technology and more innovation, as opposed to closed, patent-based monopolies.[23] Thus, advances in two key elements of the industrial revolution—steam engines and steel—both came out of a non-patented, open-sharing environment. The recent advances of Free and Open Source Software are not an anomaly but a reflection that an open model

[22] Boldrin and Levine, *Against Intellectual Monopoly*.
[23] Allen, 'Collective Invention'.

of developing knowledge is a faster and surer way to innovation than conferring state monopolies.

NATURE OF KNOWLEDGE COMMONS

The nature of the commons is obviously different if it refers to something finite rather than potentially infinite. Most of the earlier commons literature originated from goods which, though considered as public goods—for example, air—are actually finite.[24] If we keep dumping pollutants in the air, at some point its capacity will saturate; the same is not true of knowledge. If we regard all property as either private or public, we recognise only two kinds of ownership. However, a whole range of ownership exists with property held by groups or communities. Commons allow for an expansion from private to public via different forms of community ownership—providing a variety of distinctions before private property merges with the public domain.

Software, a specifically twentieth century creation, used an eighteenth-century legal form—copyright—to impose restrictive access. The problem with this restrictive access is that it does not address the specificity of software—its generally short lifespan, nature of the work, and so on.

The free software community has deployed the same legal means—copyrighting—to subvert the copyright regime. While copyleft, or the use of a specific copyright license which allows others to use it under the same copyright conditions, may be adequate in software, this alone is not enough to combat

---

[24] A discussion on the nature of the commons can be found in Charlotte Hess and Elinor Ostrom, 'Ideas, Artifacts and Facilities: Information as A Common Pool Resource', *Law and Contemporary Problems*. Vol. 66. 10.2307/20059174, 2003.
https://scholarship.law.duke.edu/lcp/vol66/iss1/5/
Also see Charlotte Hess and Elinor Ostrom (eds.), *Understanding Knowledge as a Commons*, MIT Press, 2008. However, the focus of these works is on looking at information commons and open access to information.

intellectual property rights enclosures, particularly the patenting regime. There, either public disclosure or patenting and offering the patents under license conditions similar to free software's GNU General Public License (GPL) are both being tried. GNU is a witty recursive acronym—short for 'GNU's Not Unix', the more well-known free operating system software available from Bell Labs.

Traditionally, music or books are not considered knowledge. They would be considered artefacts, which therefore could have ownership. Copyright—the dominant form of ownership of these artefacts—originates from the concept of authorship which it protects. Copyright has two aspects: one, that it confers on the author a permanent right against distortion and appropriation through plagiarism; the other is the right to make copies. The second is a temporary monopoly which can also be bought and sold. However, the digital age brings about the possibility of an infinite number of copies without any transmission loss. Books, films and music can be freely distributed at virtually no cost. How then do we consider copyright—the right of the author to recover money from his or her creative work through a monopoly—if such an arrangement produces artificial exclusions today? If technology makes reproduction a trivial exercise, should society artificially impose the monopoly of the author? If not, how do we compensate the creativity of the artist or the writer? The creative commons license, which traces itself to the GNU Public License, attempts to address some of these issues, widening considerably the ambit of the commons.

The enclosure of the commons takes place not only in areas such as science and the arts, but also traditional knowledge. As has been repeatedly pointed out, community-based knowledge is appropriated by pharmaceutical and other companies and privatised in various forms. This pertains to biological resources nurtured by communities or their specific knowledge and practices. The struggle for protecting the rights of such communities is also a struggle for protecting traditional knowledge as commons.

The Knowledge of Science and Technology as Commons 45

These commons are not public domain but the common property of a group and therefore allow for community rights, as opposed to being considered the private property of individuals and corporations. The commons license approach has been considered by the government of Kerala for protecting traditional knowledge. The Kerala Government had attempted an intellectual property rights policy for the state where traditional knowledge would be protected using a variant of a 'commons' license. Unfortunately, under the Indian Constitution, only the Union Government has the powers to make new laws in this area, and that is where this initiative languishes till this day.[25]

PRODUCTION OF KNOWLEDGE: THE INSTITUTIONAL STRUCTURE OF SCIENCE

The impact of the privatisation of knowledge and science is changing the way science is practised. Science is no longer a collaborative and open activity aimed at creating new knowledge about nature. Rather, it has become a secretive exercise in which a patent is filed before a paper is published. Ideas are not shared as they now have commercial value. This is happening at a time when the Internet and other forms of communication have enormously multiplied the possibility of open, collaborative work.[26] Monopoly over knowledge translates into the ability to extract super profits in selling a range of goods and services, whether software or a medicine or a seed.

However, the potential of a commons approach lies not only in preventing such monopolies, but also in the production of knowledge itself. The commons licenses make up only one aspect of the larger struggle to establish shared production and

---

[25] Government of Kerala, 'Intellectual Property Rights Policy for Kerala 2008', *India Environment Portal*, 1 June 2008.
http://www.indiaenvironmentportal.org.in/content/257168/intellectual-property-rights-policy-for-kerala-2008/

[26] See the reference to the Large Hadron Collider in Chapter 1.

reproduction of knowledge. The free software movement has shown the power of the new networked structures in the creation of new knowledge and new artefacts. Never before has society possessed this ability to bring together different communities and resources. What stands in the way of liberating this enormous power of the collective to produce new knowledge, and design new artefacts, is the monopoly rights and private appropriation inherent in the neoliberal intellectual property rights order.

The earlier system for developing scientific knowledge resided primarily within the structures of higher education. The universities, colleges, and other institutions of higher learning were the centres where new advances in science were located. As these centres of education were relatively autonomous of both the state and the market, the system of generating new knowledge was not closely bound by the immediate class needs of society. This produced, within the university system, a sense of independence and self-regulation; the education given to students had a larger purpose than merely serving capital or the needs of the state. This is also why the educational system provided a place for contestation, since it was where new ideas arose not only in the various disciplines but also about society itself.

The humanist view of science and technology fitted very well into this overall structure. Science was supposed to produce new knowledge, which could then be mined by technology to produce artefacts. The role of innovation was to convert ideas into artefacts —hence the protection of patenting useful ideas embodied in the artefacts.

The transformation of this system that existed for more than a hundred years has come from two different sources. One is that science and technology are far more closely integrated than before, making the distinction of scientific knowledge from a technological advance more difficult. An advance in genetics can translate to the marketplace much more quickly than before. Electronics, computers and communications have also a similar

The Knowledge of Science and Technology as Commons   47

pace of development, drawing some of the sciences much closer to the systems of production than earlier. The second is the conversion of the university system to what is essentially a profit-making commercial enterprise under the current neoliberal order.[27] The dwindling public financing of education and the rise of corporate funding has emerged as a major threat to scientific research. Market fundamentalism is, today, profoundly altering how education takes place. Students are regarded as consumers and the university education system is structured like any other commercial enterprise that looks primarily to its bottom line. A deeper analysis of nature, which has no immediate commercial use or market, is downgraded in favour of what industry considers 'lucrative' research. Not only does this distort the larger system with long-term knowledge devalued in favour of immediate and short-term gain, it also shifts research priorities away from what society needs as a whole, to servicing the needs of those who can pay. As university research is increasingly funded by private corporations, there is a wholesale shift in research priorities. Science is no longer for advancing knowledge and the well-being of society, but almost entirely for generating profits for the educational enterprise itself.

The impact of this change is evident if we compare science as it existed decades ago with how it is now. Let us take two examples. The green revolution came out of public domain science—there was no price to be paid by the farmer to utilise its advances. Today, the gene revolution is controlled by a few private corporations such as Monsanto (now merged with Bayer to become Bayer CropSciences Ltd, but retaining its brand name) along with various pharma companies. The second example is from 1955, when Jonas Salk was asked about who owned the patent to his polio vaccine.

---

[27] 'Academic administrators increasingly refer to students as consumers and to education and research as products. They talk about branding and marketing and now spend more on lobbying in Washington than defense contractors do.' Jennifer Washburn, *University, Inc.: The Corporate Corruption of Higher Education*, Basic Books, 2005.

His reply was that the people did, an answer we are unlikely to hear from a scientist today.

The passage of the Bayh-Dole Act, 1980, in the US, is the one event that went farthest towards converting publicly funded research into privatised knowledge. Inevitably, its impact in the US has been adverse, to put it mildly. *Fortune* magazine held the Bayh-Dole Act responsible for pushing up the cost of medicine: 'Americans spent $179 billion on prescription drugs in 2003. That's up from . . . wait for it . . . $12 billion in 1980'.[28] The same article also stated that the Bayh-Dole Act had actually retarded progress in science instead of encouraging it. The discovery of new molecules, a measure of innovation in the pharmaceutical industry, has actually come down since. The Act has, however, helped a few companies, universities, and scientists to become fabulously rich, at the expense, of course, of scientific development and the common people. Unfortunately, market fundamentalists the world over are pushing ideas and measures similar to the Bayh-Dole Act, ideas which will convert educational systems into University Industrial Complexes.

SCIENCE AND OPEN MODELS

Today, the information technology sector has shown that new technologies and methodologies can be developed by cooperative communities.[29] It may be argued that this sector is unique in that the 'reproduction costs' of the 'artefacts'—software—are relatively low. But is it possible to design such approaches for other areas? Is it possible, in the field of the life sciences, to have similar cooperative communities that work together to produce new products? And to envisage ways by which artefacts can be reproduced to reach

---

[28] Leaf, 'The Law of Unintended Consequences'.
[29] John Willinsky, 'The Unacknowledged Convergence of Open Source, Open Access, and Open Science', *First Monday*, Vol. 10, No. 8, 2005. https://firstmonday.org/ojs/index.php/fm/article/view/1265/1185

The Knowledge of Science and Technology as Commons 49

the community without such 'reproduction' involving high costs? To address these questions, we need to examine what structures of knowledge production are in consonance with the needs of producing new knowledge and innovation in specific sectors. Two such examples follow.

There is little doubt that genetically engineered plants are going to have an enormous impact on agriculture in the future. That they have not done so till date is due to various reasons. One, of course, is that genetically modified organisms are still in their infancy. The second and perhaps even more important point is that unlike the green revolution that came out of public domain science, the gene revolution is emerging from private domain science. The battle over the gene revolution is not the simple bureaucratic one of which department 'owns' the gene revolution, the Department of Biotechnology or the Ministry of Environment;[30] but a deeper one of which classes benefit from the gene revolution today.

The battle is intimately connected to who owns the property rights—the patents, seeds, and the chemicals—that seem to be a part of the package in the case of Monsanto and Bt cotton. In a paper written for the Food and Agriculture Organisation, Greg Traxler shows the rapid increase of transgenic crops in some countries and for specific crops: 'In 1996, approximately 2.8 million hectares were planted to transgenic crops or genetically modified organisms (GMO) in six countries.'[31] In 2019, according to the International Service for the Acquisition of Agri-biotech Applications (ISAAA), 190.4 million hectares of biotech crops

---

[30] Aniket Aga converts this inter departmental clash to a clash of epistemologies, but sometimes bureaucratic clashes are simply bureaucratic turf wars. Aniket Aga, 'Environment and Its Forms of Knowledge: The Regulation of Genetically Modified Crops in India', *Journal of Developing Societies*, Vol. 37, 2, pp. 167-183, 2021.
doi:10.1177/0169796X211001235

[31] Greg Traxler, 'The Economic Impacts of Biotechnology-Based Technological Innovations', *The Food and Agriculture Organization of the United Nations ESA Working Paper No. 04-08*, 2004.
https://www.fao.org/3/ae063t/ae063t.pdf

were planted in 29 countries.³² One could say farmers are voting with their ploughs (or tractors as the case may be) in favour of GM. The prospect of any country's agriculture passing into the hands of a few multinational companies is not a reassuring one. Even worse, most of the successful biotech seed companies are either chemical companies such as Monsanto or Du Pont, or pharmaceutical companies such as Novartis and Bayer. And when it comes to public good, the track record of both sectors has been poor. The discomfort in many countries at seeing their agriculture pass into multinational hands is not surprising.

In India, for example, the Bt brinjal debate has featured scientific worries relating to genetically modified crops for food consumption. There has been a split among scientists as well. One section has argued for far more stringent 'screening' mechanisms and controlled trials to show that GM crops are not harmful. As in the current vaccine debate, there is also a set of figures who are considered to be outside the larger scientific consensus on the possible risks from GM crops, but have an outsized influence on the anti-GMO groups. As also happens with the anti-vaccine campaign, no argument will convince those who are not willing to be convinced. In the GM debate, genuine criticism of a specific technology fix gets caught in the larger anti-GMO arguments.

Transgenic crops contain one or more genes inserted artificially either from an unrelated plant or from a different species. One of the premises in opposing GM is that transfer of genetic material from different species is not found in nature. This is not true. There are enough examples to show that it is quite common in nature—for example, in grasses (ancestors of wheat, barley, maize), lateral gene transfers are quite common and have been so from the hoary past.³³

---

[32] ISAAA in 2021: Accomplishment Report. https://www.isaaa.org/resources/publications/annualreport/2021/default.asp

[33] Samuel G.S. Hibdige, Pauline Raimondeau, Pascal-Antoine Christin, Luke T. Dunning, 'Widespread Lateral Gene Transfer Among Grasses', *New*

In my view, the critical issue is who controls global/Indian agriculture, and hence, who controls food security in countries like India. While there cannot be a technological fix to the problems of Indian agriculture, technology—and therefore GM—will still be part of the solution.[34] What safeguards society requires for GM products is still a matter of debate, but one on 'what GM' and not 'why GM'. And a debate on who benefits from the GM product: society as a whole, the farmers, or the consumers? Or the agribusiness that promotes a particular variety of seeds and specific products for private profit? For example, Bt can protect the cotton crop (or any other crop) so long as it is afflicted by the pests that Bt provides protection against. But it leads to evolutionary struggle between superior pests and newer Bt seeds. On the other hand, a genetically modified drought-resistant variety of a food crop or a GM rice modified for saline soil does not face such evolutionary pressure. The first benefits the private seed monopolies, the second the farmers. Unfortunately, with the weakening of institutions which had pioneered the green revolution, such as the Indian Council of Agricultural Research and the Indian Agricultural Research Institutes, increasingly Indian agriculture is falling under the sway of MNCs.

In order to develop useful crop varieties in the agribiotech sector, research should be conducted in multiple areas including the old-fashioned plant breeding that led to the green revolution. Instead, the bulk of 'innovative technology' in this arena currently appears focussed on making genetically modified crops (GMOs, so to say), a technology that is patent-protected by the MNC sector. Even where it is not—in countries like India that do not allow life-form patenting—keeping control over the seeds and

---

Phytologist, 230: 2474-2486, 2021.
https://nph.onlinelibrary.wiley.com/doi/full/10.1111/nph.17328

[34] Prabir Purkayastha, Satyajit Rath, 'Bt Brinjal: Need to Refocus the Debate', *Economic & Political Weekly*, Vol. XLV, No. 20, 2010.
https://www.epw.in/journal/2010/20/perspectives/bt-brinjal-need-refocus-debate.html

releasing only hybrid seeds in the market helps preserve the monopoly of companies like Monsanto.[35] An interesting step away from this corporate model of agribiotech development has been the establishment of an 'open source biology' platform, centred around new microbes useful for making transgenic plants.[36] This means bypassing Bt as the vector for transferring genetic material into other plants and therefore the Monsanto patent/s. The most advanced initiative of this kind is the Australia-based CAMBIA/BIOS. While the first acronym refers to the broader scope of promoting biological innovation for agriculture (Centre for the Application of Modern Biology to International Agriculture), the second refers to Biological Innovation for Open Society, the specific arm of CAMBIA dedicated to open-source biology.[37] This focusses on freeing the basic technological tools of biotech for general use, so that innovation at the application level is not restricted, particularly by the biggest multinationals in the biotech sector. It promotes a protected commons license for use in this regard. It also operates a web portal, Bio Forge, similar to the Source Forge of the Open Source Software Movement. While the BIOS initiative is not identical to the free-software idea, it appears to be the most developed initiative of this kind so far.[38]

[35] This and other issues relating to GM are dealt with in our article in EPW, referenced in footnote 34.
[36] Wim Broothaerts, Heidi J. Mitchell, Brian Weir, Sarah Kaines, Leon M. A. Smith, Wei Yang, Jorge E. Mayer, Carolina Roa-Rodríguez, Richard A. Jefferson, 'Gene Transfer to Plants by Diverse Species of Bacteria', *Nature* 10; 433 (7026): 629-33, 2005.
https://doi.org/10.1038/nature03309
[37] The difference between the open software and free software movements relates to copyright. In science and technology, the key issue is property rights because of patents. Open Source Biology finds alternate ways of creating similar products to patented ones, bypassing existing patents. It uses patents in the same way that copyleft licenses do—creating patented and non-patented technology that allow use by others who agree to the same principles of responsible sharing, a 'protected commons'.
https://cambia.org/bios-landing/bios-biological-open-source-licenses-and-mtas/
[38] T. Jayaraman, 'Note on Promotion of Open-Source Biology in India', Private

The Knowledge of Science and Technology as Commons

One alternate possibility being discussed globally is to take advantage of our growing ability to map the entire genetic sequence of individual organisms at much lower costs. In traditional plant breeding for advantageous traits, such a step will allow breeding programs to overcome some of the major obstacles to creating good crop varieties that also breed true—such that farmers can keep a part of their produce as seeds (unlike hybrid seeds, which need to be bought afresh from agri-business companies in every sowing season). It would then enable the identification of combinations of genes that confer a particular trait and thus allow for a reliable selection of varieties with combinations of many advantageous traits. Such a program would be of little interest to the profit sector since farmers can re-use the seed. It would require little by way of a manufacturing intermediary, since experimentally generated seed can simply be handed out to farmers and bred by them. This program would demand a large-scale cooperative global effort between plant breeders and scientists.

OPEN SOURCE DRUG DISCOVERY

A similar possibility exists in the area of drug discovery. In 1995, the Trade-Related Aspects of Intellectual Property Rights (TRIPS) Agreement introduced a uniform and higher level of patent protection across the globe. The promise that this would lead to higher levels of innovation remains a mirage. Globally, the number of such new chemical entities (NCEs) has gone down over the last few decades. Further, of the NCEs approved for marketing, a very small fraction—less than 3 per cent— constitutes a significant advance over the prevailing therapies.

An overwhelming majority of new products address the

Circulation, 2007. CSIR started a very active Open Source Drug Discovery roject under Samir Brahmachari, the then Director General of CSIR, identifying TB as its immediate target. It drew a lot of attention worldwide. After Brahmachari's retirement, the project seems to have lost steam.

needs of wealthy populations in the global North, while the disease burden falls largely to the share of the global South. While industry researches drugs for the lifestyle conditions of the affluent—obesity, erectile dysfunction, baldness, etc.—conditions such as tuberculosis, kala azar, sleeping sickness, have to make do with decades-old therapies.[39] Can open-source drug research and development, using principles pioneered by the highly successful free software movement, help revive the industry? As the cost of genome sequencing drops and the speed at which the sequencing can be done increases exponentially, it is possible to harness this power to solve the problems of health in radically different ways.

An open source model to promote innovation is not new. Used extensively in the software sector today, it organises exchange among researchers across the globe, who draw from a pooled source of information to which they pledge to plough back the new developments that accrue. Two decades back such a model might have appeared a utopia. Not so today when powerful tools are available to create virtual models that can sequence the genetic codes of humans and identify potential targets for intervention within the code.[40] It is possible to process genomic information and to create, on a much larger scale, public databases of genomic information and protein structures, identifying promising protein targets and delivering such compounds for clinical trials. It would be based on a collaborative, transparent process of biomedical development, in order to take on the health challenges that big

---

[39] For over thirty years, no new TB drugs had been introduced into the market. With Multi Drug Resistant (MDR) TB strains spreading globally, two new drugs for Tuberculosis—Bedaquiline and Delamanid—have recently been introduced to address MDR TB. See Richa Chintan, 'World Tuberculosis Day: The Patent and PIL Fight in India for Accessible Drugs', *Newsclick*, 24 March 2021.
https://www.newsclick.in/World-Tuberculosis-Day-The-Patent-PIL-Fight-India-Accessible-Drugs

[40] Bernard Munos, 'Can Open-Source R&D Reinvigorate Drug Research?' *Nature Reviews Drug Discovery*, Vol. 5, pp. 723–729, 2006.
https://doi.org/10.1038/nrd2131

pharmaceutical corporations have neglected in favour of what they perceive as 'blockbuster drugs'. A number of interesting initiatives are currently underway, on diseases from tuberculosis to malaria. One such initiative is the Council of Scientific and Industrial Research (CSIR), India, undertaking a highly ambitious program using an open-source model to generate the next generation of drugs for TB, still the number one killer in India.[41]

Malaria is another field in which a similar initiative is underway. The Medicines for Malaria Venture (MMV) has a number of projects, some of which are in the final stage of drug development or undergoing clinical trials.[42] Such a model can identify new candidates at a fraction of the cost that Big Pharma claims to spend on drug discovery. It has been argued that the major cost in drug development relates to clinical trials that need to satisfy drug regulatory agencies. Today, Big Pharma outsources clinical trials to a dispersed set of contract research organisations. A collaborative open source model could use the same route, with the difference that the entire endeavour—from the selection of promising candidates to marketing approval—is organised and overseen by a publicly funded entity or group that promises to place such research in the public domain, without insisting on patent monopolies. Or by using patents in commons similar to what open source biology uses.

Various public-private partnership initiatives are also underway. They have shown that it is possible to bring down the cost of drug discovery from the $500 million claimed by Big

---

[41] Samir K. Brahmachari, et al, 'Open Source Drug Discovery—A New Paradigm of Collaborative Research in Tuberculosis Drug Development', *Tuberculosis*, Vol. 91, 5, pp. 479-486, 2011. The project seems to have run out of steam after Prof Samir K. Brahmachari retired as DG, CSIR. https://doi.org/10.1016/j.tube.2011.06.004

[42] Instead of an open source model, MMV focusses on what it calls a Product Development Partnership, an example of public-private partnership. The public part provides the pipeline for discovery, the private part the pipeline to the market including drug trials. Medicine for Malaria Venture. https://www.mmv.org/

Pharma to less than 50 million—a drop by orders of magnitude.[43] This price advantage in developing drugs has now forced the use of such models for what are termed as 'neglected diseases' or the diseases of the poor, even though private parties still control the product pipeline to the market. This is the private part of the partnership model, in which either public or philanthropic money is used for the drug development pipeline and the market pipeline is given to the private players. As philanthropic money comes from tax savings, it is also largely public money, though it is privately controlled. It is clear that the commons approach has emerged not as a marginal view but as a rapidly emerging alternative to the current patent-ridden approach to science. The rear guard battle of global capital led by the Gates Foundation, and other similar entities is responding with hybrid approaches that allow research to be done collaboratively across groups but with its output still privatised through patents. This is similar to the Bayh-Dole Act privatising the university sector's research outputs, except that this time it works for the philanthropic sector. It is time we base ourselves not on a stronger (more restrictive) form of intellectual property rights regime, but on a larger commons approach. This direction is not only in consonance with the well-being of the people, but is also one that benefits science.

---

[43] Bernard H. Munos, 'Can Open-Source R&D Reinvigorate Drug Research?'

3. The COVID Pandemic Experience: Who Won, Who Lost?

THE COVID-19 PANDEMIC AND OUR CHOICES: SHARE KNOWLEDGE OR CHOOSE DEATH

The COVID-19 pandemic brings out all the contradictions in the society we inhabit today. We have demonstrated the ability to move in less than 12 months—the shortest time ever—from identifying the new virus to developing, testing, and producing a new vaccine at scale. And yet, after more than two years, about 30 per cent of the world has not received even one dose of the vaccine, a figure that rises to nearly 70 per cent in low-income countries.[1] We have the ability to manufacture vaccines, diagnostics, and anti-COVID-19 medicines anywhere in the world; yet intellectual property rights and WTO provisions obstruct us from doing so. We have the ability to feed all the world's population, yet see large scale malnutrition and hunger. This is the crux of capitalism: privileging the greed of billionaires over the needs of people, even when a mere fraction of this wealth would wipe out all hunger and disease.

Nothing shows up the contradiction between the greed of capital and the needs of humanity more sharply than how we have handled the COVID-19 pandemic. It lays bare all the contradictions at the core of capitalism as a world system today.

The COVID-19 pandemic has also taught us what science and technology can achieve if harnessed by society towards specific tasks. Three instances stand out, as specific responses to the pandemic. One is how quickly the genome of SARS-CoV-2 (the

---

[1] See https://ourworldindata.org/covid-vaccinations, July 18, 2023.

virus that causes the COVID-19 disease) was sequenced, followed in a matter of days by the first diagnostic kit for the virus, and, mere months later, the first set of vaccines for public use. This speed is the result of advances in science and technology.

The identification of a possible novel virus took place towards the end of December 2019. The WHO's China office was informed of cases of pneumonia with unknown cause detected in Wuhan. Science researchers in China sequenced the COVID-19 virus, the SARS-CoV-2 gene, within a week and uploaded the genome sequence to public databases on 12 January 2020, making it widely available.[2] Within days, Dr Christian Drosten of the German Centre for Infection Research at Charité, University of Berlin, created a diagnostic kit to detect the virus and shared it with the WHO. From identification to diagnostics, this was warp-speed indeed, without which the public health measures for pandemic control—such as isolation of patients, lockdowns, and physical distancing—would not have occurred.

Developing a vaccine for SARS-CoV-2 was not the only challenge. It had to be tested through clinical trials in the middle of a pandemic and also produced at scale for quick deployment. The fastest a vaccine had ever been developed before the COVID-19 pandemic was the mumps vaccine in the 1960s, which took about four years. But the two instances are not really comparable, as mumps was a well-known disease, and earlier vaccines, though not effective (possibly even dangerous), had existed since the 1940s. In this case, the entire sequence—from identification of the virus to vaccine development and deployment was completed in just 12 months. Within another nine months, a number of successful vaccines, using three major technology platforms,[3] had been delivered in more than seven billion doses.

[2] 'Novel Coronavirus (2019 n-CoV) Situation Report—1', *World Health Organization*, 21 January 2020.
https://www.who.int/docs/default-source/coronaviruse/situation-reports/20200121-sitrep-1-2019-ncov.pdf?sfvrsn=20a99c10_4

[3] Apart from the two vaccines (BioNTech-Pfizer, Moderna) using the

The two Chinese vaccines—CoronaVac and Sinopharm—along with Oxford-AstraZeneca, and Pfizer-BioNTech were the major suppliers, followed by Moderna, Sputnik, and Bharat Biotech.[4]

The flip side of the pandemic is the number of firsts it has scored on the other column: what the world did wrong. The false assurance that epidemics no longer threatened rich countries caught their public health systems unprepared for a pandemic. Their public systems (or private health care system, in the US) failed to address the challenges of such an epidemic. The pandemic also showed up the perils of the so-called globalised economy with its fragile supply chains.

Masks for preventing air-borne infections, as we know SARS-CoV-2 is, proved to be more controversial. Neither the Centres for Disease Control and Prevention (CDC) in the US nor the WHO fully understood the mechanism of transmission of the virus through aerosols.[5] It took them more than a year to issue clear directions on masking. Unlike the US CDC, countries in East Asia that had already faced the earlier SARS epidemic—China, Japan, South Korea—issued clear guidelines on masking.

---

new mRNA platform, the three other vaccines (CoronaVac, Sinopharm, Covaxin) use the older inactivated virus technology. The third set (Oxford-AstraZeneca, Sputnik) uses another platform—virus as a vector—to carry a viral protein. These were the initial successes, but more vaccines are now available, including four Cuban ones.

[4] As we now know, all vaccine immunity wanes with time and with new variants. Smriti Malapally, 'China's COVID Vaccines Have Been Crucial—Now Immunity is Waning', *Nature,* 14 October 2021.
https://www.nature.com/articles/d41586-021-02796-w

[5] Megan Molteni describes the confusion between physicists and physicians on how the terms aerosol is used. To physicians, any infectious particle smaller than 5 microns is an aerosol; anything bigger is a droplet and travels only a few feet. Physicists saw it differently; virus particles could travel in droplets much further than a few feet, as physicians believed. Megan Molteni, 'The 60-Year-Old Scientific Screwup That Helped COVID Kill', *Wired,* 13 May 2021.
https://www.wired.com/story/the-teeny-tiny-scientific-screwup-that-helped-covid-kill/

Finally, in the World Trade Organisation (WTO),[6] rich countries have blocked almost any possibility of providing affordable vaccines, antivirals, and diagnostics to the poor countries. A few days before the 12th WTO Ministerial in Geneva, UNAID's Executive Director Winnie Byanyima had said, 'In a pandemic, sharing technology is life or death, and we are choosing death'. At the Geneva Ministerial of the WTO, which ended on 17 June 2022, the rich countries did precisely that. After two years that the WTO spent in 'postponing'—or blocking—the India-South Africa proposal for a waiver on patents for COVID-19 vaccines and medicines, the club of rich countries—the European Union, the US and the UK—ensured that no worthwhile patent waiver measure was passed. The profits of Big Pharma once again trumped the lives and health of the people, just as they had done during the AIDS epidemic. The Geneva 12th WTO Ministerial chose a path that allows Pfizer and others to make huge monopoly profits at the expense of people's lives.

Let us set this in perspective: Pfizer profits have roughly doubled during the pandemic, COVID-19 vaccine patents composing a significant part of its sales.[7] If Pfizer were a country, its earnings in 2021–22—$81 billion—would place it ahead of countries such as Ethiopia, Ghana, and Kenya. These profits are generated by its monopoly over knowledge, in this case how to make vaccines for the people—despite the fact that public domain science created this knowledge in the first place. Apart from vaccines, there is also the monopoly over diagnostics and antiviral drugs, which again pushes up costs for the people while generating windfall profits for Big Pharma. The Geneva Ministerial also kicked the ball six months down the line on patents of diagnostics and antivirals. There is little chance that the block of rich countries

---

[6] 'WTO Decides: No TRIPS Waiver' *People's Health Dispatch*, 17 June 2022. https://peoplesdispatch.org/2022/06/17/wto-decides-no-trips-waiver/

[7] They are not even Pfizer inventions but those of BioNTech, Germany, which was funded largely by the German government. Pfizer secured worldwide rights from BioNtech to produce the vaccines, securing global monopoly.

will have a change of heart in six months, when they did not over the two-year course of a pandemic that has already killed millions. They will not change their minds even if the continuing COVID-19 pandemic leads to new SARS-CoV-2 variants that threaten their own populations.

VACCINES, ANTIVIRALS, AND INTELLECTUAL PROPERTY

Introduced in the early nineteenth century with mass vaccination programs along with public health systems, vaccines were our first line of defence against infectious diseases. Antibiotics emerged much later, in the 1940s and 1950s, leading to the rise of big pharmaceutical companies. While modern medicines are important in fighting diseases, public health measures including vaccines, sanitation, clean water, combined with higher incomes and adequate standards of living have done far more to reduce the disease burden in affluent countries. The socialist countries also created strong public health systems and a science and technology infrastructure for fighting diseases, which remain in place in China, Cuba, and Vietnam.

One of the consequences of the belief that infectious disease was no longer a concern of the rich was the rapid drying up of research funds needed to develop new medicines for such diseases. Annually, tuberculosis kills 1.5 million people (WHO's 2019 Global Tuberculosis Report);[8] India alone accounting for nearly half a million deaths.[9] Yet, it took nearly 40 years and the new strains of Multi Drug Resistant TB before we had new drugs entering the market.[10] For malaria, which annually infects more than 200

---

[8] Global Tuberculosis Programme, 'Global Tuberculosis Report', *World Health Organization* ISBN: 9789241565714 WHO/CDS/TB/2019.15. 2019. https://www.who.int/publications/i/item/9789241565714

[9] World Health Organization Report. 'Country Profiles for 30 High TB Burden Countries', 15 October 2019. https://medbox.org/pdf/5e148832db60a2044c2d5eca

[10] Mehdi Mirsaeidi, 'After 40 years, New Medicine for Combating TB', *Int J*

million, the last three medicines (mefloquine, halofantrine, and artemisinin) were developed 50 years ago.[11] Two out of these three (mefloquine and halofantrine) were developed by the US Army to protect its soldiers during the colonial war against Vietnamese liberation forces.[12] After the fall of the socialist bloc, Big Capital saw health and medicines as an avenue for making super profits at the expense of people's health. This led to drug companies spearheading the move for a global product patent system. Under US pressure, India and other countries conceded to the demand for a global product patent system, leading to the Trade-Related Aspects of Intellectual Property Rights (TRIPS) Agreement of 1994 and the current WTO architecture overseen by the World Bank and International Monetary Fund.

Under Bank-Fund policies, most countries in the global South have seen market-based healthcare take over the major part of their health systems, with consequences that are visible in the current epidemic. Under the Narendra Modi government, the drive towards a more privatised health care was combined with further slashing of the already low health budget allocations in India.

Public health yields 'profits' for society but not for drug companies, or Big Pharma. Profits for capital come from ill health. Amit Sengupta, one of the founding members of the global people's health movement,[13] wrote, 'Unethical behaviour of health

Mycobacteriol, 2012.
https://www.ncbi.nlm.nih.gov/pmc/articles/PMC4103685/

[11] Harin Karunajeewa, 'Mefloquine, an Antimalarial Drug Made to Win Wars' *The Conversation*, 27 April 2016.
https://theconversation.com/weekly-dose-mefloquine-an-antimalarial-drug-made-to-win-wars-55566

[12] Lynn W. Kitchen, David W. Vaughn, Donald R. Skillman, 'Role of US Military Research Programs in the Development of US Food and Drug Administration-Approved Antimalarial Drugs', *Clinical Infectious Diseases*, Vol. 43, 1, pp. 67–71, 2006.
https://academic.oup.com/cid/article/43/1/67/310038

[13] People's Health Movement.
https://phmovement.org/

care providers is directly linked with the fact that if care is linked to profit, more ill health means more profit! . . . . Governments, not markets, can ensure that health systems address the needs of the poorest and the most marginalised.' Powerful sections in the US equate public health with what they call 'socialised medicine'. It includes the American Medical Association (AMA), private insurance companies and Big Pharma. As early as 1939, AMA had argued, 'all forms of security, compulsory security, even against old age and unemployment . . . is a weakening of national calibre, a definite step toward either communism or totalitarianism'. Issuing this statement, Dr Morris Fishbein, president of AMA, warned the American public against the prospect of 'peasant medicine' and 'medical Soviets'.[14] The attack continued into the 1960s, with AMA's 'Operation Coffee Cup' against Medicaid and Medicare—the red-baiting campaign against 'socialised medicine' that launched Ronald Reagan's political career.[15] This is the core of the US right-wing view of healthcare and a major reason why the US has a dysfunctional health system. While the US has the highest per capita expenditure on health among the advanced countries, it lags well behind other advanced economies on most health indicators.[16]

[14] Richard (RJ) Eskow, '"Operation Coffee Cup": Reagan, the AMA, And the First 'Viral Marketing' Campaign . . . Against Medicare' *Huffpost*, 6 December 2017.
https://www.huffpost.com/entry/operation-coffeecup-reaga_b_45444
[15] Jacqueline Fitzgibbon, 'How Operation Coffee Cup Undermined US Universal Healthcare', *Raidió Teilifís Éireann, Ireland*, 20 October 2020.
https://www.rte.ie/brainstorm/2020/1020/1172670-operation-coffee-cup-socialized-medicine-ronald-reagan-donald-trump-joe-biden-united-states/
[16] Sarah Kliff, '8 Facts that Explain What's Wrong with American Health Care', *Vox*, 20 January 2015.
https://www.vox.com/2014/9/2/6089693/health-care-facts-whats-wrong-american-insurance
Anders Aslund, 'US Health Care is an Ongoing Miserable Failure', *The Hill*, 5 January 2019.
https://thehill.com/opinion/healthcare/423865-us-health-care-is-an-ongoing-miserable-failure/
Eric C. Schneider, Dana O. Sarnak, David Squires, Arnav Shah, Michelle M. Doty, 'International Comparison Reflects Flaws and Opportunities for

Not surprisingly, it witnessed one of the worst outcomes in the world during the COVID-19 pandemic. Even Western Europe, where a once strong public health system has been weakened by neoliberal policies, saw its hospitals,[17] even crematoriums,[18] in near collapse during the first phase of the COVID-19 epidemic.[19]

The AIDS epidemic was the first major outbreak in the West of a new infectious disease, a challenge to the presumption that the age of infections was over for them. The antiretrovirals developed for treating AIDS cost $10,000–15,000 for a year's medicine, as they were patent protected and the monopoly of a few Big Pharma companies. People continued to die for ten years as patented AIDS medicine was available only at exorbitant prices. Treatment was far beyond the reach of most patients in Africa, Latin America, and Asia. Arranging for medicine would have cost some countries more than their GDP. Finally, it was Indian patent laws—that until 2004 did not allow for product patents—which helped people to get AIDS medicine at less than a dollar a day, or at $350 for a year's supply.[20]

This did not happen without a bitter fight, as Big Pharma, backed by the US, went into action. The Pharmaceutical

---

Better U.S. Health Care', *The Commonwealth Fund Mirror*. 14 July 2017. https://www.commonwealthfund.org/publications/fund-reports/2017/jul/mirror-mirror-2017-international-comparison-reflects-flaws-and

[17] 'Hospitals in Spain and Italy on the Brink of Collapse as COVID-19 Deaths Mount', *Democracy Now*, 27 March 2020. https://www.democracynow.org/2020/3/27/headlines/hospitals_in_spain_and_italy_on_the_brink_of_collapse_as_covid_19_deaths_mount

[18] Tim Hume, 'Italy's Coronavirus Death Toll is So High That One City's Crematorium Can't Keep Up', *Vice News*, 19 March 2020. https://www.vice.com/en/article/3a8y4n/italys-coronavirus-death-toll-is-so-high-that-one-citys-crematorium-cant-keep-up

[19] See Peter Mertens, *They Have Forgotten Us: The Working Class, Care and the Looming Crisis*, LeftWord Books, 2021.

[20] Lisa Goldapple, 'India's Robin Hood of Drugs: Cipla is an Indian Drug Company Literally Saving Millions of Lives Through its Reverse-Engineered, Low-Cost Medicines', *Breakthrough*, 19 September 2016. http://breakthrough.unglobalcompact.org/briefs/cipla-indias-robin-hood-of-drugs-yusuf-hamied/

Manufacturers Association of South Africa and 39 pharmaceutical companies, including some of the world's largest and most powerful, sued the South African government in 1997 for making changes to the drug laws that would allow it to import cheap generic AIDS drugs from India and Brazil. At that time, one in seven new cases of AIDS took place in South Africa. Big Pharma wanted to make an *example* of South Africa so that other desperate countries in sub-Saharan Africa, who were equally affected by AIDS, would not follow suit. A huge global pushback by AIDS and health activists led to Big Pharma withdrawing the case.[21] Finally, the 2001 Doha Ministerial of the WTO settled the issue, allowing countries facing epidemic or public health emergencies to issue compulsory licenses for medicines to suppliers like India. Even if the procedure was cumbersome, the Doha Declaration opened the way for poorer countries to import lifesaving generic AIDS drugs on a large scale and save their people.

THE COVID-19 PANDEMIC, MEDICINES AND VACCINES

We are yet to find a magic bullet for treating the SARS-CoV-2 infection. Antivirals such as Remdesivir (Gilead Sciences), Molnupiravir (Merck), and Paxlovid (Pfizer) have shown some successes, but only in the early stage of infection. Anti-inflammatory drugs such as dexamethasone, other corticosteroids, and monoclonal antibodies work in the phase when the disease causes lung inflammation due to the body's immune system attacking the lungs. However, a number of monoclonal antibodies that worked against the Alpha and Delta variants, do not seem to work against the Omicron variant.

While dexamethasone is off-patent and available at a reasonable price, Remdesivir is patented by Gilead and sold under

---

[21] Chris McGreal, 'Shamed and Humiliated—The Drugs Firms Back Down', *The Guardian*, 19 April 2001.
https://www.theguardian.com/uk/2001/apr/19/highereducation.world

the brand name Veklury. In the US, it is priced at $3,000 for a full course, against an estimated cost of manufacture of only $10.[22] Even though Gilead has licensed Remdesivir to other companies including companies in India (Cipla, Hetero, Mylan and Jubilant), it is priced at about Rs 5,000 for a full course. This for a medicine which is easy to manufacture and should not cost more than Rs 750 for a full course, or less than one-sixth its current market-price. Molnupiravir is relatively cheaper in India, with a five-day course costing about Rs 1,400. A consortium of Indian companies led by Dr Reddy's Laboratory has got a non-exclusive license to manufacture Molnupiravir for over 100 low- and middle-income countries.[23] Even this price for Molnupiravir, to treat what may initially appear as a mild cold or a throat infection, is high for most people. If it was available at lower costs—like paracetamol or aspirin—we would see a large off-take of antivirals, cutting down on community transmission. Again, these costs would demand reducing significantly the fees on patents that Merck and Pfizer charge, hence the demand in the WTO to waive patents on drugs during the global pandemic.

From a number of vaccine trials, the WHO lists 10 vaccines that have received its emergency-use approval and a large number that are in phase three of clinical trials.[24] Apart from this list, Cuba has developed a number of vaccines,[25] three of which are already

---

[22] 'ICER Presents Alternative Pricing Models for Remdesivir as a Treatment for COVID-19', *Institute for Clinical and Economic Review*, 1 May 2020. https://icer.org/news-insights/press-releases/alternative_pricing_models_for_remdesivir/

[23] Viswanath Pilla, 'Dr Reddy's Launches Molnupiravir at Rs 35 per pill, Matching Mankind's Price', *The Economic Times*, 4 January 2022. https://economictimes.indiatimes.com/industry/healthcare/biotech/pharmaceuticals/dr-reddys-launches-molnupiravir-at-35-per-pill-matching-mankinds-price/articleshow/88689954.cms

[24] 'COVID-19 Advice for the Public: Getting Vaccinated', *WHO*, 13 April 2022. https://www.who.int/emergencies/diseases/novel-coronavirus-2019/covid-19-vaccines/advice

[25] COVID-19 Vaccine Tracker, Updated 17 June 2022.

in use there and in a few other countries. Cuba has one of the world's highest percentage of people vaccinated. Unfortunately, immunity through past infections or via vaccines does not last for long. We will require booster doses at a frequency which is yet to be determined. That means, in order to control COVID-19, we will need to have a continuous and worldwide program of COVID-19 vaccine booster doses, with stronger public health systems whose tasks include monitoring for new variants.

Why is immunising the global population important? Simply put, the more people the SARS-CoV-2 virus infects, the more chances of new variants emerging. There is a comfortable feeling among some people that the more the virus mutates, the more benign it is likely to become. This used to be a common viewpoint among a section of the medical community. However, evolutionary biologists argue that there is no evidence to show that case fatalities from viruses decrease with time.[26] Even if it is true in the 'long-run', a biological long run can be very long indeed. As John Maynard Keynes, the economist, so memorably put it, 'In the long run, we are all dead!' We are living with a pandemic that till recently involved more than 700,000 new infections every day. We are dealing with the possibility of a new variant emerging, one just as transmissible as Omicron and leading, possibly, to greater case fatalities. The transmissibility of the virus is maximum when the infected patient has only mild symptoms, is physically and socially mobile, and can therefore infect others. This is the window in which the virus spreads. Whether the patient subsequently recovers or dies has little impact on the replication of the disease. It may have an impact on our social behaviour, but that has little to do with the virus becoming more benign with time. The problems of long COVID-19 and repeat infections should preclude treating

---

https://covid19.trackvaccines.org/country/cuba/

[26] Jemma L. Geoghegan and Edward C. Holmes, 'The Phylogenomics of Evolving Virus Virulence', *Nature Reviews Genetics* 19, pp. 756–769. 2018. https://www.nature.com/articles/s41576-018-0055-5

it like any other disease that nature will tame for us.

Depending on the virus turning benign or a mythical herd immunity cannot be an answer to the current pandemic. Vaccines are crucial to any public health response as they cut down the number of new infections and, therefore, the roots of new transmissions. And yes, for the foreseeable future, we will have to live with repeating our vaccine booster doses as we fine-tune the vaccines to newer variants.

VACCINE APARTHEID, WTO, AND TRIPS

In March 2021, Tedros Ghebreyesus, the WHO's Director-General, called the distribution of vaccines 'grotesque'. He said that while some countries are vaccinating their entire populations, others have no shots for even health workers, older people, and other at-risk groups. 'The inequitable distribution of vaccines is not just a moral outrage, it's also economically and epidemiologically self-defeating'.[27] Let us not call reserving the bulk of vaccines for a handful of rich ex-colonial or settler-colonial states 'vaccine nationalism'. Let us call it what E. Tendayi Achiume, the UN special rapporteur on contemporary forms of racism called it: a *system* of *vaccine apartheid* at a global level. Writing a letter to the WTO's 12th Ministerial, Achiume pointed out, 'As of June 2022, 72.09 per cent of people in high income countries had been vaccinated with at least one dose of the COVID-19 vaccine, whereas only 17.94 per cent of people in low-income countries have been vaccinated'. The difference is stark: the percentage of vaccinated people in rich countries is three times that of poor countries even when we have the production capacity to vaccinate the entire target population. In the US, toddlers are being vaccinated, while in Africa only 18

---

[27] 'Inequity of COVID-19 Vaccines Grows "More Grotesque Every Day"—WHO Chief', *UN News*, 22 March 2021.
https://news.un.org/en/story/2021/03/1087992#:~:text=%E2%80%9CIn%20January%2C%20I%20said%20that,at%20a%20regular%20news%20briefing

per cent—or one in five persons—had, till June 2022, received the required two shots.

The figures were even starker in November 2021: 65 per cent of the people in high-income countries had been vaccinated with at least one shot, and only 3 per cent in Africa.[28] This at a time when the rich countries had stored a billion doses in their refrigerators instead of sharing them with the world.[29] Without China, which has emerged as the largest producer and exporter of COVID-19 vaccines, Africa, Asia, and Latin America would have been much worse off.[30] There was even a smear campaign against Chinese vaccines, calling them not good enough, their effect waning with time. As we now know, in time all vaccines wane in their ability to stop new infections, particularly faced with newer variants of the virus. Why is it that when the scientific community could deliver vaccines against a new virus so quickly, and not one but a number of them, we failed to make and distribute them cheaply across the globe?

Apart from patents and know-how, what obstructed a quick ramping up of global vaccine production was that the rich countries—the US, the EU, and the UK—used a crude me-first model of vaccine manufacture, blocking the export of supplies of intermediate products and raw materials. The US used a 1950 (Korean war-vintage) Defence Production Act for this purpose.[31]

[28] Nurith Aizenman, 'Low Income Nations Need COVID Vaccines. Rich Countries Have Millions of Unused Doses', *NPR*, 8 November 2021. https://www.npr.org/2021/11/08/1053647185/low-income-nations-need-covid-vaccines-rich-countries-have-millions-of-unused-do?t=1636455627997

[29] 'Coronavirus, The Billion Doses in the Fridge That Doomed Africa', *Instituto Humanitas Unisinos (IHU) Magazine*, 29 November 2021. https://www.ihu.unisinos.br/614846-coronavírus-aquele-bilhão-de-doses-na-geladeira-que-condenou-a-áfrica

[30] Xiaoyi Wang, 'China, Not COVAX, Led Vaccine Exports to the World's Middle Income Countries in 2021', *Health Policy Watch*, 10 February 2022. https://healthpolicy-watch.news/china-covax-led-vaccine-exports-lmic-2021/

[31] Allison Martell and Euan Rocha, 'How the U.S. Locked Up Vaccine Materials

Table 3.1: COVID-19 Vaccines and Their Funding

Vaccine	Patent Owner/Owners	Funds
AstraZeneca/ Oxford vaccine	University of Oxford initially said that it would be available to all who wanted to manufacture the vaccine, but later signed an exclusive deal with AstraZeneca. Serum Institute Pune has a non-exclusive agreement with AstraZeneca for producing the vaccine.	Public
Johnson and Johnson	US government gave $1b for developing COVID-19 vaccine. BARDA gave another $454m.	Public
Moderna	Moderna and NIH. National Institute of Allergy and Infectious Diseases researchers made the key advances, though Moderna claims that some of the key patents are theirs. Received $1b from US government in research funding.	Public
Pfizer/ BioNTech	BioNTech developed the technology through a grant of euro 375m from German Federal Ministry of Education and Research. Pfizer's role was in clinical trials and commercialising the vaccines. US government gave Pfizer an order of $1.95b for the first 100m doses.	Public
Sinopharm	Sinopharm, as per its website, is directly under the state-owned Assets Supervision and Administration Commission (SASAC) of the State Council (Sinopharm website)	Public
Sinovac	SinoVac is a private company. Chinese government contributed an estimated 1b renminbi—about $140m—in developing CoronaVac, the vaccine.	Public
COVAXIN	Technology was developed by Indian Council of Medical Research (ICMR) and given to Bharat Biotech and National Institute Virology (Pune). Given to Bharat Biotech for vaccine production.	Public
Novavax/ Covovax/ Nuvaxo-vid	Received $1.6b from Operation WarpSpeed of the US Government and $384m from CEPI set up by Bill and Melinda Gates Foundation.	Largely public
CanSino/ CONVIDECIA	Funds not clear; presumably government supported.	

Thus India's export commitment regarding vaccines was adversely affected on two counts: one, by the US restrictions, and two, by the need to address the devastating second wave in India.

This is where capital, intellectual property rights—patents and trade secrets—come in. If we trace the original research, or the knowledge on which these vaccines were based, they almost all came from public domain science funded by different states. As Table 3.1 shows, the research and initial advances for production received from the state made it possible for private players to make a killing with their vaccines. Had the vaccines failed, the state (or states) would have borne the losses. The governing logic of a successful private-public partnership is that it privatises the profits, and where it is unsuccessful, the public bears the losses. Table 3.1 also shows how virtually the entire knowledge base of the vaccines, whether for the newer vaccine platforms, the mRNA or the adenovirus vector platforms, came almost entirely out of public domain science but was patented by private players with either the open or covert support of the state.

If we combine the direct R&D funding provided by governments with Advance Purchase Agreement commitments, 80–90 per cent of the funding of successful COVID-19 vaccines is from public funds. A research group in Geneva—the Knowledge Network on Innovation and Access to Medicines—has detailed the public funds used in COVID-19 vaccine development, which gives us these figures.[32] As the money for purchasing vaccines is given before any vaccine has been developed, it is treated by researchers as public money. The putative total sum is an underestimate as the Chinese Sinovac is incorrectly shown as private investment, which it is not. The role of philanthropic capital is particularly interesting. The Bill and Melinda Gates Foundation has played

---

Other Nations Urgently Need', *Reuters*, 7 May 2021. https://www.reuters.com/business/healthcare-pharmaceuticals/how-us-locked-up-vaccine-materials-other-nations-urgently-need-2021-05-07/

[32] 'COVID-19 Vaccine R&D Investments', *Knowledge Portalia*, 8 July 2021.

an outsized role in determining global health policies. Emerging as a major funder for the WHO (the second-highest contributor to its budget), the Gates Foundation virtually dictates policy prescription to the world body. The advanced capitalist countries were aware that another fiasco like that of AIDS, where the rich countries had fought tooth and nail against any relaxation of their monopoly over drugs, would cost them dearly.

It is here that Bill Gates, who built his trillion-dollar empire on Microsoft's intellectual property (IP), plays an outsized role in COVID-19 vaccines and medicines. In global health policy, he has taken up the mantle of protecting the intellectual property rights of the drug companies. The Oxford team had made a public declaration that their research effort for a vaccine, if successful, would be available to any company wanting to make the vaccine. It was Gates and his foundations that persuaded the Jenner Institute, Oxford University, to give the patents instead to AstraZeneca as a monopoly.[33] Governments and the WHO are hamstrung by lack of funds, leaving the field clear for the Coalition for Epidemic Preparedness Innovations (CEPI) and the Global Alliance for Vaccines and Immunisation (GAVI): two vaccine initiatives in the role that WHO and governments should be playing. Both are controlled by the Bill and Melinda Gates Foundation.

One of the reasons for the WHO's lack of funds is the US government's effort to cut its funding, even withdrawing from the WHO during Trump's presidency.

People might argue that if private philanthropic money comes into public health, it is after all private money and can do what it wants. The catch here is that it is not really private money, its source is the large tax exemptions that capitalist countries provide to philanthropy. This helps capital channel money in a direction of

---

[33] Jay Hancock, 'They Pledged to Donate Rights to their COVID Vaccine, Then Sold Them to Pharma', *KFF Health News,* 25 August 2020. https://khn.org/news/rather-than-give-away-its-covid-vaccine-oxford-makes-a-deal-with-drugmaker/

its choosing. Capital gets very large tax exemptions, retains control over the money and can even use it to capture global and national policy.

Speaking to Sky TV UK, on India and South Africa's proposal in the WTO for lifting intellectual property protection on COVID-19 vaccines and medicines, Gates claimed that IP is not the issue: 'Moving a vaccine . . . into a factory in India, . . . it's only because of our grants and expertise that can happen at all'.[34] This is a variant of the white man's burden that underpinned the West's colonial enterprise. The intervention of Gates and others provides a rehash of the AIDS debate, where Western governments and their industry—the Big Pharma—argued that developing generic AIDS drugs would lead to poor quality drugs and theft of Western intellectual property.

Indian companies are the largest manufacturers in the world by *volume* for existing vaccines.[35] The global MNCs or Big Pharma's share is much larger when it comes to *money:* their patent-protected vaccines with monopoly pricing make more money. For example, the Serum Institute in 2020 supplied 28 per cent by volume but received only 3 per cent by price. Contrast this with GlaxoSmithKline (GSK) who supplied 11 per cent by volume and received 40 per cent by price. GSK, Pfizer, Merck, and Sanofi controlled 90 per cent of the global vaccine market by price while supplying less than 30 per cent by volume (See Figure 6.1 in Section II). The model that the West and philanthropic capital sells is simple: let Big Pharma make the big bucks even if

---

[34] 'COVID-19: Bill Gates Hopeful World "Completely Back to Normal" by end of 2022 - and Vaccine Sharing to Ramp Up', *Sky News*, 25 April 2021. https://news.sky.com/story/covid-19-bill-gates-hopeful-world-completely-back-to-normal-by-end-of-2022-and-vaccine-sharing-to-ramp-up-12285840

[35] WHO Team (Immunization, Vaccines and Biologicals) 'Global Vaccine Market Report', Working Draft, *World Health Organization*, 30 November 2020. https://www.who.int/publications/m/item/2020-who-global-vaccine-market-report

it bankrupts the poorer countries. Western philanthropic money will 'help' the poor Third World to get some vaccines, if slowly, as long as the West gets to call the shots. This is the old AIDS model, of 'subsidised' AIDS drugs with USAID and philanthropic money, which Africa, Asia, and Latin America had rejected at the time.

If we do not address the intellectual property rights issue in this pandemic, we are likely to see a repeat of the AIDS tragedy.[36] Most countries have compulsory licensing provisions that will allow them to break patents in case of epidemics or health emergencies. Even the World Trade Organisation (WTO) accepted in its Doha Declaration (2001) after a bitter fight, that countries in a health emergency have the right to allow any company to manufacture a patented drug, or even import it from other countries. But the rich countries ensured that the compulsory licensing procedures were complex and cumbersome. The Doha Declaration only covered patents, and not trade secrets that are important in vaccine manufacture. Countries are also afraid to break patents, even if there are provisions that allow for it in their laws and in the TRIPS Agreement. It is their fear of a bullying US and its reprisals. Under the Domestic Trade Act, the US issues Special Reports—USTR 301—threatening any country with trade sanctions that tries to compulsorily license a product. India figures prominently each year,[37] for daring to issue a compulsory license in 2012 to Natco Pharma for Nexavar, a cancer drug, which Bayer AG was selling for $65,000 a year.[38] Marijn Dekkers, the CEO of Bayer, was widely quoted saying that what India had done was 'theft': 'We did

---

[36] Goldapple, 'India's Robin Hood of Drugs'.
[37] Reji K. Joseph, 'US Tries To Thwart TRIPS Flexibilities in The Midst Of a Pandemic', *Livemint*, 1 May 2020.
https://www.livemint.com/opinion/online-views/us-tries-to-thwart-trips-flexibilities-in-the-midst-of-a-pandemic-11588320583933.html
[38] 'India's First Compulsory Licence Granted to Natco for Bayer's Cancer Drug' *The Hindu Businessline*, 14 November 2017.
https://www.thehindubusinessline.com/companies/Indias-first-compulsory-licence-granted-to-Natco-for-Bayers-cancer-drug/article20408026.ece

not develop this medicine for Indians . . . . We developed it for Western patients who can afford it'.[39]

WTO'S 12TH MINISTERIAL: GREED AND DEATH TRIUMPHS OVER LIFE

The idea of sharing patents and supplying COVID-19 drugs and vaccines at concessional rates to poor countries was backed by all in the World Health Assembly, WHA-73, barring the US and its loyal camp followers, the UK and the EU.[40] In October 2020, India and South Africa moved the WTO for a temporary waiver of patent rights for vaccines, diagnostics and medicines during the COVID-19 pandemic.[41] This had overwhelming support among countries and public health groups all over the world. Under Trump the US had opposed any relaxation of TRIPS provisions even under the conditions of a pandemic. The Biden administration reversed this stand but considerably narrowed down the scope of the waiver to *just patents*. This was contrary to the original proposal: to waive *all intellectual property rights on COVID-19 vaccines, diagnostics, medicine, including industrial designs, copyright and trade secrets*. These were required to take vaccines from research and development to *scaled up* production at an industrial scale. Without extending the waiver to other

[39] James Love, 'Bayer CEO Marijn Dekkers Explains: Nexavar Cancer Drug is for "Western Patients Who Can Afford It,"' *Knowledge Ecology International*, 23 January 2014.
https://www.keionline.org/22401

[40] Sarah Boseley, 'US and UK "Lead Push against Global Patent Pool for COVID-19 drugs,"' *The Guardian*, 17 May 2020.
https://www.theguardian.com/world/2020/may/17/us-and-uk-lead-push-against-global-patent-pool-for-covid-19-drugs

[41] 'Waiver from Certain Provisions of the TRIPS Agreement for the Prevention, Containment and Treatment of COVID-19: Communication from India and South Africa', *World Trade Organisation* IP/C/W/669(20-6725), pp. 1-4, 2 October 2020.
https://docs.wto.org/dol2fe/Pages/SS/directdoc.aspx?filename=q:/IP/C/W669.pdf

property rights, the Biden stance of waiving only patent rights is more optics than real change.

Even at the level of optics, why did the US suddenly change its position? The US has been relatively isolated in its America First policy of hoarding vaccines and vaccinating all people in the United States before exporting to the rest of the world.[42] China has emerged as a major source of vaccines for large parts of Africa, Asia, and Latin America. This endangers the Biden plans of a grand alliance against China to isolate it globally. The US geostrategic view is that the Western big Pharma companies should dominate the rich countries' markets and that of the global rich more generally, these being the only two markets of interest to Big Pharma. Moderna is slated to make a profit of $19 billion this year (2021), Pfizer $13 billion. This is the market that the rich countries want to protect.

For the rest, the US was banking on its new Quad partner India to provide vaccines to the rest of the world through the WHO's Covax program. The Covax program, though nominally the WHO's, is dominated by Bill Gates and his various vaccine initiatives: Bill and Melinda Gates Foundation, GAVI, and CEPI. The Serum Institute of India with AstraZeneca and Novavax vaccines, and Biological E with the Johnson & Johnson (Janssen) vaccine, would make available 2.6–3 billion doses per year from India, helping to vaccinate the global population by 2022–23.

This strategy faltered due to the failure of the Modi government to use India's indigenous capability for ramping up its vaccine production. It did not invest in the indigenous vaccine and biopharmaceutical capacity. Instead, the Modi government believed in the 'magic' of the free market that would automatically provide all the vaccines needed. It also did not anticipate the devastating

---

[42] Jon Cohen, 'Unveiling "Warp Speed", the White House's America-First Push for a Coronavirus Vaccine', *Science*, 12 May 2020.
https://www.science.org/content/article/unveiling-warp-speed-white-house-s-america-first-push-coronavirus-vaccine

second wave, which led to India banning exports till it had met its own needs.

The proponents of the patent monopoly—Big Pharma and philanthropic capital—argue that a patent waiver is useless as it is a lack of technology, knowledge, and capital, not patents, that is holding up vaccine production outside the rich countries. If patents were not stopping vaccine production in other countries, why did Big Pharma and the rich countries oppose such a waiver in the WTO?

According to Big Pharma, a patent waiver on vaccines will dis-incentivise research and is a huge blow to those who innovate. What they hide—and this is not new—is that most of the research money for the new vaccines has come from public funds. A *Lancet* paper by Olivier Wouters and others shows that governments and philanthropic organisations have *given more than 10 billion dollars* for the development of the current crop of vaccines and to various promising vaccine candidates.[43] This does not include the billions of dollars that the US and the UK governments paid to Pfizer and AstraZeneca in advance orders. The argument of providing a monopoly to Big Pharma for incentivising drug discovery is therefore a bogus one. The money and most of the research comes from public funds and government laboratories. The role of philanthropic money in developing private monopolies merits a more detailed discussion. But, for this chapter, it should be treated on par with public money as it comes out of tax-free dollars.

There are three major technology platforms that have emerged in the current lot of successful vaccines:

1. The old-fashioned inactivated virus as a vaccine, such as Sinovac, Sinopharm, and ICMR-Bharat Biotech;

---

[43] Olivier J. Wouters, Kenneth C. Shadlen, Maximilian Salcher-Konrad, Andrew J. Pollard, Heidi J. Larson, Yot Teerawattananon, Mark Jit, 'Challenges in Ensuring Global Access to COVID-19 Vaccines: Production, Affordability, Allocation, And Deployment', *Lancet*, 13 March 2021. https://pubmed.ncbi.nlm.nih.gov/33587887/

2. The relatively innocuous virus as a vector to carry a SARS-CoV-2 protein, as done by AstraZeneca, Cansino, and Gamaleya Institute's Sputnik V; a variant of this is directly injecting a protein sub-unit as a vaccine; and
3. The mRNA that tells the body's cells to produce the SARS-CoV-2 protein, e.g., Pfizer-BioNTech, Moderna.

All of these three technology platforms have produced successful vaccines. Almost all the Big Pharma arguments on why patent waivers are not much use, are about mRNA vaccine platforms. The mRNA vaccines are not of immediate public health interest to most countries; these require an ultra-cold chain, otherwise they degrade rapidly. The cost and effort involved in building such an ultra-cold chain preclude the use of mRNA vaccines in mass vaccination programs in most countries.

Of interest to most countries are the inactivated virus vaccines, the adenovirus vector vaccines (or its variant, the protein sub-unit vaccines). While the inactivated virus vaccines have been around for a long time and are produced by a number of countries, scaling up their production requires money, equipment, and time. Comparatively, scaling up adenovirus vectors to carry the protein fragment or scaling up the production of a sub-unit of the viral proteins is easier. But it comes with the constraint that more trade secrets are associated with the production of viral vector or protein sub-unit vaccines than with the older inactivated vaccines.

India, China, and South Korea, the three largest generic vaccine manufacturing countries, definitely have manufacturing capability for biologics, for either adenovirus vector vaccines or for protein sub-unit vaccines. Zydus Cadila, in partnership with India's Department of Biotechnology, has an indigenously developed vaccine which has been granted emergency use authorisation in India. China's CanSino has developed an adenovirus vector vaccine that has been granted authorisation by the Chinese authorities and also the WHO. A similar vaccine is among those Cuba has developed.

WHO's Covax platform, directed by GAVI and CEPI where Bill Gates has an outsized influence, focussed much more on the new vaccine platforms—the mRNA and the adenovirus vector vaccine platform of AstraZeneca (sourced from Serum Institute India), also Johnson & Johnson—and not the traditional inactivated virus vaccines. Covax had committed to supply 2 billion vaccines by December-end 2021, and managed to supply less than half of that. This forced many countries in Africa and Latin America to start their own initiatives, striking direct deals with Chinese manufacturers who stepped into the breach.

There were various conceptual issues with the Covax platform. True to the ideology of philanthropic capital, it was designed primarily as a vaccine allocation-cum-delivery platform, leaving out issues of manufacturing, technology transfers and building a vaccine infrastructure. When pharma companies' deliveries failed, or were diverted to supply more lucrative markets, Covax did not have much to offer, besides begging the rich countries who had hoarded extra shots to part with them before the expiry date.

It was only after the failure of the Serum Institute of India and Johnson & Johnson to fulfil their contracts that WHO-Covax agreed to use inactivated virus vaccines from China's Sinopharm and Sinovac, and India's Covaxin. But even now, the bulk of the WHO's Covax supplies use the new vaccine platforms. Dr Ricardo Palacios of the Butantan Institute, São Paulo, pointed out at a South Centre event on 1 April 2021 that CEPI and Covax mostly funded the newer vaccine technologies and disregarded the inactivated virus vaccines. The latter are effective, cost less and can be produced easily in many developing countries. As of June 2022, China's Sinovac is the world's leading COVID-19 vaccine supplier. Before dismissing these vaccines as yesterday's technology, let us note that this is still *the vaccine platform for all flu vaccinations* across the world with about 1.5 billion doses administered per year. Covax's 2022 performance is better as we have surplus vaccines today, beyond what the world requires. But the yawning gap between rich and poor countries still remains, with rich countries supplying fourth booster shots and even vaccinating toddlers, while the entire continent of Africa has vaccinated just about half its population.[44]

For any company involved in biologics, adenovirus vector or protein sub-unit vaccines are pretty much routine technology. India has about twenty such companies. So have South Korea, China and Bangladesh. There are also countries in South East Asia, Latin America, and Africa which have biologic drug manufacturing capability and can become major manufacturers. Cuba has developed five vaccines, of which four have undergone trials successfully and two are already in use. China's Cansino and India's Zydus Cadila have developed a protein sub-unit vaccine. To argue as Bill Gates did on Sky TV, 'without us teaching the Indians' (or Koreans, Chinese, Latins, Africans, Arabs, Cubans) 'or our money, they cannot produce vaccines', is just a repetition of the 'white man's burden' that cloaked the earlier colonial enterprise.[45] Even with the supposedly most advanced mRNA vaccines, Médecins Sans Frontières has identified 100 companies that can manufacture them if the patents and know-how are shared.[46] This is the key bottleneck. C-TAP, the WHO's technology sharing pool, has been boycotted by almost all leading manufacturers of COVID-19 vaccines. It had been presented as an alternative to the

---

[44] Even this figure does not tell the full story. If we take out outliers like Morocco, Egypt, and few others, the rest of Africa coverage would be less than 10 per cent.
[45] 'COVID-19: Bill Gates Hopeful', *Sky News*.
[46] Achal Prabhala and Alain Alsalhani, 'Pharmaceutical Manufacturers Across Asia, Africa and Latin America with the Technical Requirements and Quality Standards To Manufacture mRNA Vaccines' *Access IBSA*, 10 December 2021.
https://accessibsa.org/mrna/

demand that WTO/TRIPS should waive all intellectual property rights. If so, it definitely has not worked.

The other mechanism favoured by Big Pharma is patent pooling: the pharmaceutical companies license a set of manufacturers for a patented product but retain control of the process and pricing. It is meant to ensure that an AIDS-like backlash does not lead to the compulsory licensing (by countries) of their patented product. It keeps the patents under Big Pharma control, lowering prices for the poor countries whilst reducing the kind of public anger against exorbitant prices that had broken out over the AIDS drugs. The WHO, rich countries, and philanthropic capital have all backed the patent pool against the India-South Africa initiative in the WTO, which has been backed by more than 100 countries.

IN WTO'S GENEVA MINISTERIAL, DEATH TRUMPS LIFE

The UNAIDs Executive Director, Winnie Byanyima, had appealed for patent waivers before the 12th Ministerial of the WTO in Geneva, saying that the world would otherwise face a grim future. It is worth quoting Byanyima again: 'In a pandemic, sharing technology is life or death, and we are choosing death'. As we have seen, rich countries at the 12th Ministerial did precisely that. No worthwhile patent waiver measure was passed. While countries in need of vaccines could now issue compulsory licenses more easily than they could under the provisions of the Doha Accord, the new procedure actually made it more difficult for countries in a position to supply vaccines—such as China and India—to do so. To whom do countries issue compulsory licenses if not those with manufacturing capacity?

In vaccine manufacture, it is not the 'formula' of the vaccine that matters. Unlike with many medicines which are small chemical molecules and therefore easy to patent, vaccines are large molecules and belong to what are called biologics. The key to their manufacture is not the 'formula' of the compound but the process of replicating complex large molecules accurately while manufacturing at an industrial scale. This is the 'know-how' involved, which is guarded not under patents but under trade secrets. These trade secrets can be duplicated or secured by somebody who knows the process. However, companies who do so open themselves to costly legal action, including in the WTO. As also the threat of unilateral sanctions by the US, the EU, and the UK.

The upshot is that Pfizer and other Big Pharma companies will go on making huge windfall profits at the expense of people's lives, even if new SARS-CoV-2 variants emerge as the pandemic continues. It is why less than 20 per cent of the African continent, with a population of 1.21 billion, had been fully vaccinated as of 15 August 2022, while vaccine doses went abegging in the United States.[47] We have the vaccine production capacity to immunise the entire global population. But that is not in the interest of Big Pharma, to whom profits matter far more than human lives. Even when countless lives might be saved and the possibility of new, dangerous variants reduced. The only waiver in the Geneva Ministerial was on compulsory licenses for vaccines. It did not address patents on diagnostics and antiviral drugs. It also did not include in its waiver the other property rights, for example, trade secrets, essential to the mass production of vaccines.[48] While the patents on antivirals as a cure for COVID-19 are important, and will certainly cut down the deaths and complications of long COVID, again, patents come in the way of their use. The antivirals are effective only in a small window of the first few days of the disease, and that means they have to be available *cheaply* so that people can buy them from the chemist. Control over patents and

[47] https://africacdc.org/covid-19-vaccination/
[48] Médecins Sans Frontières, 'Lack of a Real IP Waiver on COVID-19 Tools is a Disappointing Failure for People', Statement, 17 June 2022.
https://www.msf.org/lack-real-ip-waiver-covid-19-tools-disappointing-failure-people

The Covid Pandemic Experience

the high prices of these drugs inhibit the market for them. A small market and high prices lead to a catch-22 situation: the prices are high because the market is small; the market is small since the prices are high!

Again, the open licensing of antivirals could make it possible for a large market to emerge. But this is what the WTO does not allow. The route of compulsory licensing under WTO is cumbersome, and its 'relaxation' in the Geneva Ministerial means that countries like India, which were crucial in fighting the AIDS epidemic, are supposed to opt out as suppliers. They cannot become the antiviral drug suppliers for COVID-19 that they were for AIDS.

Why don't countries that have the capacity to manufacture advanced vaccines—India, China, Russia, South Africa—come together and offer technology and supplies to the rest of the world? Why don't countries collaborate with Cuba, a biologic powerhouse, to produce vaccines locally? As mentioned earlier, Cuba has developed four such vaccines, two of which are already in large-scale production.

The answer lies in the 'rules-based international order' propagated by the club of the rich. The rules include sanctions on many countries, including Russia, Cuba, and China. For those not yet under sanctions, there is the threat of future sanctions from the same gang of three—the US, EU, and UK—who teamed up to defeat the India-South Africa initiative in the WTO. In addition, and this bears repeating, the US also has its domestic law, USTR 301, to 'protect' its intellectual property, under which it threatens countries with sanctions. Even courts in the US hold that its domestic laws trump international law. India and China figure prominently every year in the list of countries whose laws and actions do not conform to US domestic laws. If the US and its allies do not win in the WTO, they use *their* 'rules-based order' where *they* get to make the rules.

Welcome to our brave new world, where to paraphrase Winnie Byanyima, death triumphs over life.

The challenge we face as global citizens is this: even if there are medicines and vaccines against COVID-19, will they be available to everybody at prices that poor people and poor countries can afford? This is a battle that individuals or countries cannot win by themselves; we either stand united or we perish individually. Only collective action and global unity, or what Ronald Reagan called the dreaded 'socialised medicine', can save the world from the COVID-19 pandemic. Only a sharing world can defeat the COVID-19 pandemic. This what global capital dreads, and what we need to fight for.

Section II

Paradigm Shifts in Technology

4. Understanding the Philosophy of Technology[1]

Perhaps technologists have been too busy changing nature through their practice to theorise much about the nature of their practice. And the world of philosophers has not been very kind to technology, regarding it as a poor country cousin to science, not quite in the same league. To understand the relation between science and technology and society, it is necessary to understand what technology is, its inner laws, its relation to both the realm of knowledge at large and to its parts, the meaning of technological progress, and a host of other matters that are essentially philosophical questions.

Carl Mitcham and Robert Mackey point out that 'the philosophy of technology, like philosophy itself, deals with second-order theoretical questions, such as: Is technology mere applied science? What is the meaning of technological efficiency?'[2] Henryk Skolimowski takes a more restrictive view of the philosophy of technology and says that it 'belongs to the realm of epistemological enquiry and attempts to situate technology within the scope of human knowledge'.[3] The question is not one of the relation of technology as one branch of knowledge to all the other branches, but a deeper one about what place technology occupies in the realm of things, the place of the objects it produces in the real world. It is in the absence of coherent and consistent

---

[1] As mentioned in the Introduction, this chapter is an edited version of what I wrote in 1978. Four decades later, these views continue to be relevant.
[2] Carl Mitcham and Robert Mackey (eds.), 'Introduction: Technology as a Philosophical Problem' in *Philosophy and Technology: Readings in the Philosophical Problems of Technology*, Free Press, 1972.
[3] Henryk Skolimowski, 'The Structure of Thinking in Technology' *Technology and Culture*, Vol. 7, 3, pp. 371–83, 1966.
https://doi.org/10.2307/3101935

answers that technology is in danger of being robbed of its own identity, its soul.

The spectacular growth of science and technology in the twentieth century, and their consequent interaction and interpenetration has tended to blur the distinction between technology and science, at least to the layperson. In the academic world, this has been interpreted as the conquest of technology by science and its conversion to applied science. Of course, the more perceptive have observed that there seems to be an intrinsic difference between science and technology: the former is general, easily transferred, supranational, while the latter is specific, primarily national, and difficult to transfer.[4] Science diffuses easily; technology is slow to percolate through national and cultural borders. Nevertheless, a theoretical framework that would account for these differences is yet to be constructed.

Before we get into what technology is, it is necessary to understand what it is not. There are two closely related misconceptions that must be cleared away, the first being that technology is applied science. To some, the mere application of the laws of science can lead to the production of material goods. At a more sophisticated level, the standpoint Mario Bunge takes is to say that (i) technology has rules; and (ii) these rules can be deduced from the laws of science with suitable empirical modification, depending on the level at which they have to operate.[5] His example: quantum laws may be completely unnecessary when dealing with the macro level. Bunge's schema

---

[4] Derek de Solla Price, 'The Difference Between Science and Technology', *The International Edison Birthday Celebration Lecture*, 1968.

[5] Mario Bunge, 'Technology as Applied Science', Technology and Culture, Vol. 7, 3, pp. 329–47, 1966.
https://doi.org/10.2307/3101932
The title of his original contribution was 'Towards a Philosophy of Technology', but as the name of the symposium and the subsequent issue 'stole' the name of his paper, he was persuaded to change the title to 'Technology as Applied Science'. Carl Mitcham, *Thinking Through Technology—The Path Between Engineering and Philosophy*, University of Chicago Press, 1994.

Figure 4.1: Technology Rules and Application of Science

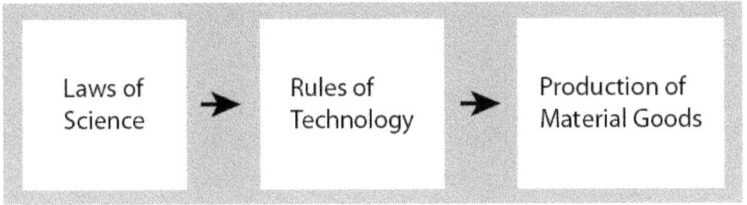

Source: Mario Bunge, 'Technology as Applied Science', Technology and Culture, Vol. 7, 3, pp. 329-47, 1966.

may be represented as in Figure 4.1.

However, there is more to technology than either the laws of science or rules derived from such laws. The application of the laws of gravity and the laws of fluid mechanics will produce suicide if one jumps from the twentieth floor of a building. But applying the same laws in another manner, one can overcome the constraints of nature to achieve flight. The fundamental differences between these two applications should be obvious. The question is not of the mere application of science, but one of recognising the particular manner of its application.

The second misconception is that while science is believed to be quantitative and mathematical, technology is a complex of techniques and intuition: in other words, 'technology is skills'.[6] Here again, the question as to what skills constitute technology is not made clear. Art involves skills, but not the same ones as technology. This approach does not 'furnish clear conceptual distinctions between techniques [of] acting and skills of making, nor between technological and other skills'.[7] The two misconceptions are not mutually exclusive. Indeed a juxtaposition of these two views occurs in many works—where the authors think that while in the past technology was a complex of skills, now, with the growth of

---

[6] James K. Feibleman, 'Technology as Skills', *Technology and Culture*, Vol. 7, 3, pp. 318-328, 1966.
https://doi.org/10.2307/3101931

[7] Mitcham and Mackey, *Philosophy and Technology*.

science, technology is turning into applied science. What they fail to recognise is that however much science may develop, in order to create new artefacts, technology will still have to bridge the gap between what is known and what is not.

Common to both these approaches is the belief that technology is the application of certain rules, either based on science or on experience and skill. But the central question is not whether technology is based on knowledge (both scientific and intuitive); rather, how it amounts to a distinct enterprise. It produces artefacts that serve a specific material function—a functionality incorporated in the artefact produced. The product of technology is not objectified knowledge, but the objectification of a material function, which emerges at the prompting of social need. A compressor is designed on the basis of the laws of thermodynamics, but is not a simple reflection of these laws. Embodied in the compressor is the specific function it must fulfil—that of compressing air. This is what forms the invariant core of technology. Whatever the changes in method, or in the internal structure of technology, this core provides an unbroken continuity to such dissimilar devices as a stone-age cutting tool and a numerically-controlled machine tool of the electronic age.

We now come to the next important question: What is the distinction between science and technology? Scientific activity is ordered towards knowledge of reality via laws formulated to explain regularities in nature. Such laws go on to form a body of theoretical constructs. Technical activity produces artefacts which are based on the knowledge of external reality and incorporate either a specific function or a complex of functions. Science is meant to understand nature, technology to change it. While scientific progress is measured by how closely its theoretical constructs approximate to external reality, technological progress is measured by how effectively it fulfils a range of functions. As Henryk Skolimowski says, 'technology pursues not knowledge but effectiveness in

Understanding the Philosophy of Technology

producing objects of a given kind'.[8] Here 'effectiveness' is to be understood in a much broader sense than Skolimowski uses it. He reduces effectiveness ultimately to engineering efficiency, albeit of varying kinds in the different branches of engineering. Instead, effectiveness should be understood in the sense of socially effective, because ultimately it is society which determines whether to accept a certain product or not. A combustion engine may be more efficient than all earlier ones, but cost may prohibit its use. Thus, while scientific progress means a fuller understanding of nature, technological progress is a more effective fulfilment of a material and social need.

We have already stated that science involves understanding nature, while technology changes nature. In changing nature to our purposes, technology has to base itself not only on what we know, but also on experience and intuition in order to overcome what we do not know. Science can never know nature completely, and experience and intuition are permanent components of technological knowledge. Derek de Solla Price has noted that science is supranational, an international movement; technology, on the other hand, is national.[9] Ian Jarvie says, 'Science aims to investigate the universe and most general laws it obeys, technology works within those laws, concentrating on what is possible in narrow localities of the universe.'[10] The highly localised character of technology stems precisely from the fact that while science can be transferred and disseminated easily, experience cannot be so easily transferred. This is why technology has to be absorbed and assimilated, a process much slower than the diffusion of science.From dealing with the distinction between science and technology, it is necessary to elaborate on the connection between the two. After all, the question of distinction would have been

---

[8] Skolimowski, 'The Structure of Thinking in Technology'.
[9] Derek de Solla Price, 'The Difference Between Science and Technology'.
[10] Ian Charles Jarvie, 'Technology and the Structure of Knowledge' in Mitcham and Mackey, *Philosophy and Technology*.

unimportant if a close connection between them did not exist. In order to change nature, it is necessary to understand nature first: to understand nature, it is necessary to change nature. The basic tool of enquiry for the scientist is experimentation, i.e., changing nature. Technology provides the instruments of enquiry—whether the telescope or the microscope—which advance science.

Technology and the problems of production constantly pose new problems before the scientists. Solutions to such problems create new technologies, which in turn pose new problems.

Science and technology do not interact in a vacuum; rather, they operate within the matrix of social relations. In history, we can find long periods of stagnation of both science and technology. The most vigorous growth is seen precisely in those periods where science and technology are in close interaction. In the process of material production, human beings had to study nature in order to appropriate from it the things they needed. The birth of science originates from technology, as noted by Benjamin Farrington: 'Science, whatever its ultimate developments, has its origin in techniques, in the arts and crafts, in the various activities by which man keeps life going on, keeps body and soul together'.[11] In less resonant cadences, John D. Bernal makes the same point: 'It was in the ways of extracting and fashioning materials so that they could be used as tools to satisfy the prime needs of man that first techniques and then science arose'.[12] It is also natural that 'its flourishing periods are found to coincide with economic activity and technical advance'.[13] Technology and the problems of production provide the stimulus to the development of science. The development of geometry in ancient Greece was based on knowledge the Egyptians had derived from surveying, pyramid-building, and military needs. Similarly, Boris Hessen has shown the relation between the technical problems and problems of

---

[11] Benjamin Farrington, *Greek Science: Its Meaning for Us, Part 2*, Pelican, 1944.
[12] John D. Bernal, *Science in History*, Vol. 1. Pelican, 1954.
[13] Ibid.

production which stood behind Newton's discoveries.[14] When the unity between science and technology is ruptured, both science and technology descend into sterility. Science and technical activity are human activities: if they are separated socially, the link between science and technology can snap. This is what happened in the Middle Ages where there was practically no advance in either. Science had become the prerogative of a class of leisured people who considered it demeaning to take part in actual production. Technology was the preserve of the slave or the serf. As a result, both domains were stifled. The impact of technology is felt not only in that it supplies new problems, but new instruments of enquiry. Hitherto, the history of science has paid little attention to the history of such instruments. It is well known that the development of particle physics was possible due to the creation of good vacuum pumps from the middle to the end of the nineteenth century. Similarly, the computer and the electron microscope have opened up completely new branches of science, and given a tremendous impetus to the development of its existing branches.

The development of science is determined by the needs of production or other social needs. This is all the more true of technology. Along with the efficiency of a technical process comes the question of its economic process viability. Thus, Hero of Alexandria (c. 10–c. 70 CE) had built a steam turbine which remained unused: slave power was much cheaper than steam power. In economies where production is for the market, profitability determines whether a technology is accepted or not. In other words, technological progress is not simply increased efficiency but also determined by the cost of improvement. It is socially determined, and the history of technology is interwoven with the history of society.

Science and technology must be seen as sub-systems of a larger

---

[14] Boris M. Hessen in Nikolai I. Bukharin (eds) *Science at the Crossroads: Papers from the Second International Congress of the History of Science and Technology*, Second Edition, London, 1971.

Figure 4.2: Society, Science and Technology

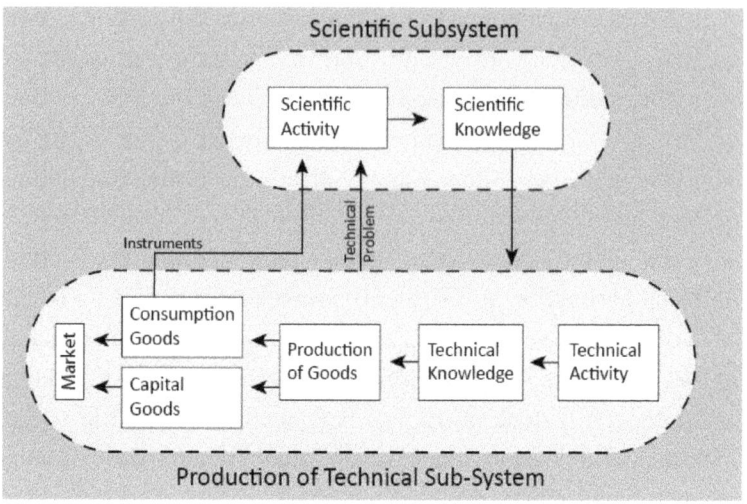

system—the society within which they interact. Diagrammatically, the whole (society) and the parts (science and technology) are shown in Figure 4.2.

In this schematic outline attempting to define technology, along with its relation to science and society in general, the historical dimension of the issue has been touched upon only briefly, and perhaps inadequately. There are other important questions that have till now received little attention—primarily regarding the structure of technological knowledge, and the methods of technology—without which no philosophy of knowledge is complete. The intention here is to pave the way towards a philosophy of technology, which can only come out of the interaction between technologists, scientists and social scientists.

5. Restoring Conceptual Independence to Technology

Technology is generally viewed with a great deal of suspicion in today's trendy environment of anti-modernity. It is held to be intrinsically violent—even genocidal—in its aim of dominating and controlling nature.[1] Common examples range from the Nazi gas chambers to the Bhopal gas tragedy—these events are perceived as somehow inherent in the vision of technology. And science is thought to share the ideology of dominance over nature that characterises technology. Science, in this sense, is a virtual synonym for technology. In short, the Hiroshima bomb and $E=MC^2$ are frozen together in time and space in such a way that their separation becomes an artificial exercise. The more 'classical' view of technology is not much kinder. The philosopher Hans Poser remarks on this view:

> Philosophers, in most cases, do not deal with such dirty things as technology. They prefer to discuss the rationality of animal rationale instead of the products of homo faber. Scientists, in most cases, look down on technology as a kind of science-less application of science; only if they need some sophisticated new measuring instruments do they accept technology as an auxiliary science.[2]

---

[1] A number of writers take the position that science and technology share a common purpose of dominating nature and are, hence, intrinsically violent. See, for instance, Shiv Vishwanathan, *A Carnival of Science*, Oxford University Press, Delhi, 1997. Some feminist critics of science and technology also use the theme of domination to attack current science and technology as male-centric.

[2] Hans Poser, 'On Structural Differences Between Science and Engineering', *Society for Philosophy and Technology*, Vol. 4, 2, pp. 81-93, 1998. https://scholar.lib.vt.edu/ejournals/SPT/v4n2/pdf/POSER.PDF
See also Klaus Kornwachs, 'A Formal Theory of Technology?' *Society for*

In other words, technology gets located as a poor second cousin of science in the knowledge domain. It is either 'applied science' or 'technoscience', with the earlier traditions of technology taken to be isolated 'techniques' and consigned to its pre-history. In this view, techniques concern the properties of nature—earlier known empirically and now replaced by the more rigorous scientific knowledge of the day. The supersession of traditional and empirical knowledge is held to define modern technology, as distinct from earlier artisanal techniques; hence the description of technology in this scheme as either 'applied science' or 'technoscience'. The progression of technique to technology transfers it from its earlier subaltern location to the more elite one of science, but it does not lose either the taint of commerce or the odour of sweat.

The primary reason for the identification of science with technology seems to be their shared skills and tools—the experimental method, and mathematical reasoning. Historically, they have also often shared people: figures such as Archimedes and Galileo are famous not only for the development of mechanics and physics, but also for developing a variety of machines. The need for defence (or offence) saw the learned men of different times harnessed to the military machine of their age, early forms of the military-industrial complex! The services Leonardo da Vinci offered to his patrons were primarily his technical abilities, not painting and sculpture. However, while the science that Archimedes or Galileo developed is still useful (for example, the Archimedes principle), their military machines, catapults, etc. appear to be more or less obsolete.[3] Hence the belief that technologies of old are more akin to crafts, with empirical knowledge standing in for

---

Philosophy and Technology, Vol. 4, 1, 1998.

[3] George Basalla, *The Evolution of Technology*, Cambridge University Press, Cambridge, 1993. Basalla has shown that older technological artefacts reappear in many of the products of the industrial revolution and therefore are present even today, albeit in new forms. However, the artefacts are no longer used for the functions for which they were originally designed and therefore can be considered obsolete.

Restoring Conceptual Independence to Technology

the real underlying principles of nature—which were not grasped. The extrapolation follows that with our better understanding of science today, technology is technoscience, distinct from the earlier version with its proximity to the crafts. This division permits philosophers of science, with their fixation on physics, to comfortably abandon the grubby world of artefacts with its location in labour and contemplate the more elevated one of ideas. To quote Hans Poser once again: 'Sciences [. .] are the coddled child of at least philosophers of science, who, up to now, have developed a paradigm of science depending on their fixation on physics'.[4] Both these views look at technology and science as inseparable and consider that the two have to be examined only in their unity; thus the issue of addressing two separate halves is effectively collapsed into one. For those who would like to attack technology as a Manichean force subverting the good in nature, the task is considerably simplified by not having to distinguish knowledge of nature from its modification. Those committed to the Eurocentric project of showing that all knowledge flows from the 'Western' tradition of enquiry can tidy away inconvenient details—such as important technological advances made by other civilisations—by filing them under 'proto-technology'.

The belief that science and modern technology are intrinsically 'Western' is a deep-seated one that comes up in both optimistic and critical accounts of scientific progress. However, while the appropriation of science for the 'West' is relatively easy, not so in the case of technology.[5] The monumental work of Joseph Needham, which showed that the compass, the printing press, and gunpowder all originated in China, made it very difficult to argue that other civilisations did not have true 'technology', although

[4] Poser, 'On Structural Differences'.
[5] James Poskett gives us a much-needed corrective to the Eurocentric view of science and reason being uniquely Western. That the Enlightenment is coterminous with colonial loot, genocide and slavery is conveniently airbrushed out of history. James Poskett, *Horizons: A Global History of Science*, Viking, 2022.

the argument has persisted that non-Occidentals—unlike the Greeks—did not possess 'true science and true mathematics'.[6] The backdoor appropriation of technology required its subsumption under science. If science is Western and technology has become technoscience, it can be argued that modern science and technology are essentially Western. I reproduce below a long quote from an article by Evandro Agazzi, to show how these views are presented as arguments, even after Martin Bernal's incisive account of their origin in racism during the late nineteenth and early twentieth centuries.[7]

> The suffix, 'ology' that we find in the word technology, invites us to take advantage of the theoretical aspect that is usually bound up with its use (compare theology, sociology, philology, ethnology); it serves to indicate the presence of some kind of 'scientific', or at least theoretical dimension.
>
> In fact, the Greek term *techne* already included this theoretical aspect, since it was used to indicate the capability of justifying, of 'knowing why' a certain efficient procedure was efficient. I would maintain that the modern concept of technology can be interpreted as a new way of expressing the conceptual content of the Greek term *techne*. Without indulging in detailed historical reconstruction, we can say that Western civilisation finds what is perhaps the most decisive element of its specificity—as regards other great civilisations in human history—in that it explicitly introduced the theoretical demand into the domain of practice and of doing. What we might well call the 'invention of the why', rising from within Hellenic civilisation in the sixth century BCE, led in that same context to the birth of both philosophy and science in the

---

[6] Joseph Needham, *The Grand Titration: Science and Society, East and West*, Allen and Unwin, London, 1969.
[7] Martin Bernal, *Black Athena: The Afroasiatic Roots of Classical Civilisation*, Vols I and II, Rutgers University Press, 1987.

strong sense. (They were originally one and the same.) The very demand which led philosophers to ask for the reasons for the existence and constitution of the cosmos (and to postulate principles and first causes to provide such an explanation), was also what moved the first mathematicians to provide the reasons (by means of demonstrations) for the properties of numbers and figures; other peoples had discovered them only empirically, translating them into practical rules of calculus. In following this impulse, it was inevitable that a search for the 'why' should eventually take up the different sorts of efficient knowing that men had used in various fields; and this gave birth to the notion of *techne*: efficient action where we know the reasons for its efficiency and that it is founded upon them.[8]

I will take the third view, that science and technology are different (if allied) human activities, with different goals and objectives. This is what Dudley Shapere has argued, that science is intuitively concerned with a theoretical understanding of the world and technology with the construction of material artefacts and structures.[9] Walter G. Vincenti has made it difficult to argue that technology is mere applied science: 'Making something that works economically, reliably and safely is a rather different thing in purpose and consequences from running a scientific laboratory experiment. Such differences help explain why engineering can never be simply applied science'.[10] Scientific activity discerns

---

[8] Evandro Agazzi, 'From Technique to Technology: The Role of Modern Science', *Society for Philosophy and Technology*, Vol. 4, 2, 1998.
https://scholar.lib.vt.edu/ejournals/SPT/v4n2/pdf/AGAZZI.PDF

[9] Dudley Shapere, 'Building on What We Have Learned: The Relations Between Science and Technology', *Society for Philosophy and Technology*, Vol. 4, 2. 1998.
https://scholar.lib.vt.edu/ejournals/SPT/v4n2/pdf/SHAPERE.PDF

[10] Walter G. Vincenti, 'The Technical Shaping of Technology: Real-World Constraints and Technical Logic in Edison's Electrical Lighting System.' *Social Studies of Science*, Vol. 25, 3, pp. 553–74, 1995.
http://www.jstor.org/stable/285506

laws, classifications or patterns in nature. These are then used for explanatory and predictive purposes. While technological activity does employ knowledge—whether law-like or empirical—its primary goal is to produce artefacts that incorporate some designed social function. The artefacts can be either material or 'code'; the computer program is an artefact in this scheme, though not a material one. The relation between science and technology may evolve over time, but this does not mean that the nature of scientific or technological activity has changed in the process. I will examine specific aspects of technology that I believe make it a distinct endeavour from that of science.

I will also explore why the interpenetration of science and technology does not invalidate the fact that they are distinct kinds of activities.

The account given above, of the nature of science and technology, does not address the currently influential stream of social constructivism.[11] I accept one of the fundamental propositions of social constructivism—that technology does not unfold in a unilinear fashion driven by some inner logic, but is also the result of a series of social choices—a view in common between the social constructivists and a much older tradition asserting that society shapes technology. The problem I have with the normative account of social constructivism is its rather contrived neutrality on the question of why certain artefacts prove successful; its principle of symmetry where the analyst remains 'impartial to the real properties of her object of analysis, viz, technology'.[12] This studied neutrality towards factors explaining the success of a technology is also to be found in social constructivist accounts of scientific theories—once again neutral on the question of why particular theories enjoy privileged status. The social constructivists have

[11] Wiebe E. Bijker, Thomas Parke Hughes and Trevor Pinch (eds.) *The Social Construction of Technological Systems: New Directions in the Sociology and History of Technology*, MIT Press, Cambridge, 1987.

[12] Philip Brey, *Philosophy of Technology Meets Social Constructivism*, Society for Philosophy and Technology, Vol. 4, Nos. 3-4, 1998.

drawn strong criticism here, the most influential critics being Langdon Winner and Walter G. Vincenti (1995).[13] Winner also criticised the social constructivists' inability to summon moral or political principles and take an evaluative stance towards specific technologies. That there are social choices being made in the development of technology does not tell us anything about the choices that *should* be made from an emancipatory viewpoint. Should we build a nuclear bomb at all? Or fight against it? Or should we be neutral on this issue and confine ourselves to asking why the scientists/technologists settled on a particular technology, size and shape for the bomb. The second problem with the social constructivists' account is the belief that technology is infinitely plastic and the design space is therefore infinite. The feasible design space for most technologies is in fact strictly limited, and social shaping drives artefacts further into already (technically) bounded choice sets.

Vincenti states his problem with this approach:

> The potential for social influence—even social determination or, to use recent scholarly parlance, 'social construction'—clearly exists at all times. Technical constraints, however, have their base at least partly in physical laws and requirements. To circumvent them is frequently impracticable and in some cases physically impossible. When social influences and technical constraints conflict, the constraints must often prevail.[14]

In other words, social choices do play a role but only once the major shaping of technology has already taken place, and they relate to the appearance of the artefact more than to the

---

[13] Langdon Winner, 'Upon Opening the Black Box and Finding it Empty: Social Constructivism and the Philosophy of Technology', *Science, Technology, & Human Values,* Vol. 18, 3, 1993 and Walter G. Vincenti. 'The Technical Shaping of Technology'.
https://www.jstor.org/stable/689726

[14] Vincenti, 'Engineering Knowledge, Type of Design, and Level of Hierarchy'.

fundamental function for which it has been designed.

I will address in the following sections the historical relation between science and technology, and its evolution. I will also discuss the role of science in the design paradigm, arguing that it acts as a part of the constraining envelope in the design of artefacts.[15] Thus, nature's laws, as explicated by science, are not only applied to design artefacts but they also limit the possibilities of their design, setting out what cannot be done. This is perhaps its larger role in technology: defining what not to do. For those who are disturbed by this relegation of science to an apparently negative role, I must point out that in developing any artefact it is crucial to delimit its possible design space. I will next address whether there are design paradigms in technology, and how they change. Klaus Kornwachs has pointed out (though his observation applies more to artefacts than design) that since the results we do not want are much more extensive than the goals we definitely want, all our technological actions are necessarily incomplete and risk being unable to avoid reaching an undesirable state (failure). Since technology cannot anticipate every roadblock ahead (or, in engineering-speak, all prevention modes), designs will occasionally fail. This brings us to an examination of the role of failure, the unknown unwanted states which surface, and which lead in turn to changes in design paradigms.

HISTORY OF SCIENCE AND TECHNOLOGY RELATIONSHIPS

A conventional view of the history of technology is that it was largely a collection of techniques and became science-based only

[15] The proposition that science cannot be used directly but only by a preventing process has been elaborated by Kornwachs, who has built upon Mario Bunge's analysis of the difference between the structure of scientific and technical knowledge. Kornwachs, 'A Formal Theory of Technology?'. Mario Bunge 'Towards a Philosophy of Technology', in Carl Mitcham and Robert Mackey (eds.), *Philosophy and Technology: Readings in the Philosophical Problems of Technology*, Free Press, New York, pp. 62–76, 1972.

after the industrial revolution. In this process, technology came to be institutionalised on the same lines as scientific institutions and broke free from its past conservative character, transcending the guild form of organisation. The earlier craftsmen who produced artefacts were judged to be divorced from the realm of scientific theories, basing themselves largely on rules of thumb and empirical knowledge. With the growth of scientific knowledge, it becomes possible to substitute laws of nature for 'crude' rules of thumb and thus generate a technology which is imprinted with science, if not fused fast with it.

A certain assumption regarding the nature of science underpins this view of technology. In this scheme, if the rule of thumb and empirical knowledge can be replaced by science, science must indeed have moved fairly close to truth. It follows that in this view of scientific progress—as closer and closer approximations of truth—the progression is not discrete but continuous. Otherwise, a closer approximation of 'truth' could well uncover a new level in science, of which the existing scientific knowledge offers no adequate account. If an artefact is developed that utilises properties at this unprecedented level, it must again fall back on empirical knowledge to explain itself, without achieving a scientific understanding of nature.

The relationship between science and technology stated above also assumes that technology was leading science earlier, while science is today in the lead. However, the implicit absence of paradigm shifts makes this conventional account of science—now lagging behind technology, now leading it—rather suspect.

The historical account asserting that technology enjoyed a lead earlier and is trailing science now, is generally assumed to be self-evident, with a few examples given more as illustrations than proof. Even these examples are contentious, and empirical evidence can be used to support different conclusions. Thus, according to Dudley Shapere, the invention of the steam engine initiated the science of thermodynamics and, in this account of the history of science, technology earlier had primacy over science.[16] However, Shapere has also quoted Donald S.L. Cardwell (1995),[17] who argues that Guericke's study of atmospheric phenomena was used in the Newcomen engine, the first commercially used steam engine. It was appropriately called the atmospheric engine. It is not relevant to my discussion whether the steam engine created the field of thermodynamics or was itself a product of atmospherics. My contention is simply that claims regarding the primacy of technology over science in the past are contestable; not only are alternate views possible but they exist.

The primacy of science over technology during the last hundred years has been taken to be axiomatically true. Undoubtedly, science and technology are much closer today as human activities than in the past. This does not automatically give either primacy or a lead to science in its relationship with technology. The example most often quoted is that electronic circuits are a result of our understanding of quantum tunnelling and may therefore be considered an example of science leading technology. Shapere states:

> Many of the innovative features introduced in the great revolution of our times, in electronics and computers, are in part applications of quantum tunnelling, a process which was literally unthinkable before quantum mechanics, and which could not have been applied in invention or even conceived, even by the cleverest of inventors before the advent of the theory.[18]

The historical record is far less clear. A characteristic of electronic components is that they are non-linear and result in non-linear circuits at the macro level. The first semiconductor device was patented by J.C. Bose well before quantum mechanics

---

[16] Shapere, 'Building on What We Have Learned'.
[17] D.S. Cardwell, *The Norton History of Technology*, New York, Norton, 1955.
[18] Shapere, 'Building on What We Have Learned'.

had developed.[19] Even if we accept the precedence of science in terms of knowing the quantum phenomena, it by no means follows that technology is able to use quantum laws to design artefacts. In fact, the quantum laws only explain individual devices. The core of electronic engineering is developing complex circuits using such devices where the behaviour of each component must be known at the quantum mechanical level and also at the macro level. We have no means of executing such designs 'scientifically' using both macro and quantum mechanical levels of laws; the intersection of the macro and quantum worlds is where all our semi-conductor devices lie, a domain that cannot be solved mathematically with our current state of knowledge. In terms of circuit design, these semiconductor devices are analogous to the old-fashioned diodes and triodes, and they behave similarly. While the use of semiconductor devices leads to miniaturisation and large-scale integration, the fundamental nature of electronic circuits does not change. They remain as non-linear circuits, due to the use of non-linear electronic components. Since we do not as yet possess the analytical tools to solve them, they are still designed using old-fashioned trial and error and rules of thumb. If there is one area in modern technology in which the technologist behaves very much like the craftsman of yore, it is in electronic engineering. So, to offer electronics as proof of either the primacy of science over technology, or the fusion of science and technology, is to misunderstand the fundamental processes at work.

The example of electronic engineering is a particularly good one as it raises two other issues crucial to understanding technology. One is the role of different forms of knowledge in technology and the other is crossing from known levels of nature into the unknown, and their impact on technology.

---

[19] Darrel T. Emerson, 'The Work of Jagadis Chandra Bose: 100 Years of Millimeter-Wave Research,' in *IEEE Transactions on Microwave Theory and Techniques*, Vol. 45, 12, pp. 2267-2273, Dec. 1997. doi: 10.1109/22.643830

CONFRONTING THE UNKNOWN

In creating artefacts, technology has to address its lack of complete knowledge of the real world. As science does not provide complete knowledge of nature in any frame of reference, how do we develop artefacts that can be predicted to work in the absence of complete knowledge? The technologist uses two methods to address this lack of knowledge: experiments to work out empirical relations; and a quantification of this lack of knowledge—whether scientific or empirical—in a factor of safety that should be described more correctly as a 'factor of ignorance'.

That technology has become far more science-based stems from the belief that there is less ignorance today than before. In other words, this belief regards nature as two-dimensional and bounded, our increasing knowledge of it progressively reducing the area of the unknown. However, if nature is not bounded and has levels, it poses serious problems for this view. As technology does not remain within the bounds of the known, it still has to address the unknown. With the crossing of levels in nature, the problem is even more acute. Technology has to push its way out of the 'envelope of the known' to confront a new level where nature may be still largely unknown.

This issue comes more readily into focus if we take the earlier example of electronic engineering. Originally, electrical circuits were linear as they used only linear elements: resistance, capacitors, and inductors. The laws of science—Ohm's law, Kirchoff's law, and other similar laws—can be used to compute and solve such circuits, and therefore to design devices containing such circuits. However, electronic circuits use non-linear elements such as Bose's two-element junctions, diodes, and semiconductor junctions. As the resulting nonlinear circuits are not amenable to analytical methods, the use of electronic devices in technology means again relying on empirical knowledge and factoring in ignorance.

Were technology confined to the domain of linear electrical

circuits, the argument that science has reduced ignorance and restricted the role of empirical knowledge to developing predictable devices would undoubtedly hold true. With the introduction of non-linear circuits and electronics, the certainty that characterised linear circuits vanishes. Unfortunately, the unknown always yawns open at the feet of the unwary. Electronics is just another example of the same—rather than proof that science has been playing tag with technology, and technology is now leading science.

Let us turn briefly to another aspect of this lead-lag view of technology, namely, the role of instruments in advancing science. Quite often, insignificant advances in technology have had a profound effect on science. This is because a new level in nature can be uncovered with even a minor technological progression. Consequently, some improvements in the instruments of enquiry open up domains or levels of reality not apparent earlier. The development of better vacuum pumps in the second half of the nineteenth century was by itself no major technological landmark, but it uncovered the quantum world, leading to the 'demise' of classical physics. Similarly, a rather minor advance in glass-grinding created the optical lens. The astronomical and microscopic worlds became accessible to scientists as a consequence. The development of the instruments of enquiry has grown today to such a level as to lead to arguments that scientific reality itself is mediated by technology and that this is the sine qua non of the scientific enterprise.[20] Using the instruments of enquiry as examples of how technology leads science, even as a number of technological artefacts use new principles of science, Shapere has argued that technology and science continuously play leapfrog with each other. I would argue that the examples given of science leading technology are based on an insufficient understanding of how technology actually develops. The idea of the primacy of science or its leading technology is really trying to

---

[20] Basalla, *The Evolution of Technology*.

find a common metric between two incommensurable domains. The philosophy of technology is quite often a derivative discourse of the philosophy of science, and shares the scientist's prejudice regarding technicians in their laboratories. Vincenti has also clarified that unlike science, which is driven from laboratories, technology is driven by laboratories *and* design offices, the latter playing an even more important role in developing technology. If technology and science are held to be different enterprises, then the question of lag-lead or the primacy of one over the other becomes meaningless. No technological artefact can be constructed without understanding nature in some sense; nor can such an artefact be based purely on established knowledge.

DESIGN PARADIGMS AND THE ROLE OF FAILURE

The crucial difference between scientific activity and technical activity is that technology has, as its starting point, an abstract design through which it fulfils a certain function. This function is not merely to meet an existing need, but often creates a new social need through novel artefacts. It must be borne in mind that no social need can be created unless there is a basis for it in society; the artefact must base itself on certain existing aspects of human or social reality even as it attempts to create new needs. The design of any artefact starts from an abstract model—however fuzzy—which incorporates the design function of the artefact. The creative process of innovation is much closer to the artistic process than conventionally granted. It is no accident that a Leonardo da Vinci offered his services to his patrons primarily as an engineer. The ability to create a mental image before developing it in practice was something in common between his sculpture, painting and design of artefacts, and made him a master in each of these fields.

The issue I would like to raise here is whether there are design paradigms in developing artefacts and whether such paradigms undergo shifts in the way paradigms do in science.

Restoring Conceptual Independence to Technology

A number of historians and philosophers have held that while science has discontinuities and revolutions, technology is far more continuous and conservative.[21] This seems particularly surprising at a time when the breaks with continuity in technology are happening at a faster rate than ever. In some measure, the basis for asserting that technology tends to be conservative and progresses without discontinuities is to shift the technological revolution to the scientific domain, and posit that technological revolutions occur due to exogenous factors. In this view, only scientific revolutions cause discontinuities in technology, as technology follows into the discontinuous domain opened up by science. But as we have already noted, this view of the development of technology is a restrictive one, particularly when the reality studied in science is itself mediated today by technology.

Any artefact must be based on a mental design. This mental design has to incorporate the specific functions the artefact will perform. The design paradigm is the construction of this artefact within the envelope of the known, and quantifying the outer limit of the unknown through a factor of ignorance. A change of design paradigm occurs when we either create a novel artefact to fulfil functions that did not exist earlier, or when we extend the scale of the artefact. In either case, the designer is straining against the envelope of the known and thus bringing out new aspects or levels of reality.

The aspect of spanning levels of reality (say, from electrical to electronic engineering) has been dealt with earlier. A change of scale that breaks out of the design envelope is not commonly realised. Change in the scale of an artefact is one of the commonest evolutionary processes in technology. The scaling up or down of products may appear to be a trivial exercise, but it can also bring to light either new properties or new levels of reality.

The scaling of artefacts may have to do with the various

---

[21] Ibid.

dimensions of the artefacts or with changes in the environment within which they operate. For our purpose, a dimensional change or a change of temperature that the artefact has to withstand, both constitute changes of scale. For example, in construction, the length of a simply supported beam can be increased till it collapses under its own weight. With short beams, their weight may not be important. However, with increasing length, the weight increases till it becomes an important variable in the problem. Thus, the scaling up of an artefact leads at some point to dramatic consequences in the domain of design, with the artefact bursting out of the envelope within which the original design paradigm was constructed. Similarly, a change of temperature beyond the design domain of the artefact can lead to the artefact failing, as happened in January 1986 with the failure of the O-Ring seals in the space shuttle *Challenger*, where very low temperature conditions caused the rubber seals to lose elasticity and break.[22]

Exceeding the envelope of design is generally not a quantitative issue. If the failure occurs because of a mistaken stress limit, changing the stress limit would be sufficient to readjust the design envelope. It is when a qualitatively new phenomenon is encountered during a scale change that we have a change in the design envelope itself. Again, to take recourse to a historical example, we might see wind effects and the resulting vibrations as a new 'dimension' in bridge design, necessitating a shift of the design paradigm. This is what happened in the famous Tacoma Narrows Bridge failure in 1940, leading to a design change in all long-span bridges.[23] It is interesting to note that recent work

[22] Trudy E. Bell and Karl Esch, 'The Fatal Flaw in Flight 511', *IEEE Spectrum*, Vol. 24, 2 February 1987.
https://doi.org/10.1109/mspec.1987.6448023

[23] While the collapse of the bridge is commonly believed to be due to resonance, and this is what we were also taught in our classrooms, the phenomenon was more complex. It collapsed due to 'self-excitation, aeroelastic instability'; or more simply, flutter! The design changes required for stopping flutter are different than for avoiding resonant frequencies. Bernard J. Feldman, 'What to Say About the Tacoma Narrows Bridge to Your Introductory Physics

treating the shift of design paradigms has focussed on studying failure in order to identify the domain of validity of the design envelope.[24] Merely studying the success of an artefact cannot be used to validate the underlying design paradigm. The failure of the artefact brings out the inadequacy of the design paradigm, and factoring this in is a necessary part of any appraisal. For example, in the Tacoma Bridge failure, not addressing the issue of flutter.

Paradigms exist in technology as in science, and paradigm shifts occur in very much the same way. Failure analysis plays a similar role in changing design paradigms, as does the Popperian notion of falsifiability in science, and can be used to validate or invalidate paradigms. While failure and falsifiability are similar in nature, there is one significant difference between the two. Falsifiability is a softer notion, while failure is far more definitive in nature. However, both are open to the kind of immunising auxiliary hypothesis that attempts to save an existing paradigm by modifying the framework only quantitatively, without addressing the need to change the framework or add a dimension to the framework that was not considered earlier.

What constitutes a design paradigm and what role does science play in it? It should be obvious that in our view the design paradigm is central to the artefact. The design paradigm incorporates the primary functions of the artefact as well as embedded knowledge of nature, both empirical and law-like. The primary feature of the artefact may also be based on a novel property discovered by the enterprise of science.

Thus, the design paradigm is a set of abstract principles that gives the idealised artefact its essential function. Science and its laws act more to constrain the design. They restrict the possible design space of the artefact and provide the outer dimension of the

---

Class', *Physics Teacher*, Vol. 41, 92, 2003. https://doi.org/10.1119/1.1542045

[24] Henry Petroski, *Design Paradigms*, Cambridge University Press, Cambridge, 1994.

feasible. This is remarkably close to the current view of complexity and biological evolution: that evolution is not about optimising life forms in a continuous domain with infinite possibilities, but optimising them within a space that has far fewer options. Taking this view of the evolution of the artefact, both the scientific constraints imposed by nature and the social choices made restrict the design space of the artefact.

Technology, neither in its evil incarnation violating nature, nor its avatar as applied science, comes anywhere close to technology as practised by technologists. The process of the evolution of artefacts must address the issues raised by various views of technology. Obviously, the framework adopted will dictate the choice of technology and its larger social role. However, I do not propose to address these issues here. There is only one observation that I wish to make: technology cannot be considered merely a tool that is used or misused. The very conception of social function in designing artefacts involves existing social realities that influence the design. However, knowledge, and more particularly the laws of science, also enter the design paradigm. In other words, though human beings make artefacts, they do not make them as they please. Contingent social and scientific 'laws' exist that also determine the success or otherwise of the artefact.

This account is an attempt to restore to technology a conceptual independence that current accounts of it seem to lack. The prevailing views of technology, either as violently subjugating nature, or as a derivative of science, seem to be unlike the way it is actually practised. A bird's-eye view of science and technology can therefore be quite misleading in defining the relation between the two. As a practising technologist, I have more of a 'worm's-eye view' of the process, and find this gives a more satisfying account of technology since it takes as its starting point a better understanding of the design process and the design paradigm. Once the design paradigm is addressed in more detail, the role of

Restoring Conceptual Independence to Technology   113

science and its relationship to technology takes on an appearance quite different to the one promoted by current accounts of the interaction between science and technology.

6. The Dynamics of Technology and Self-Reliance

The freedom movement had adopted self-reliance as a basis for India's development. With independence, India set out to act on this goal and a key step was planning its economy. It also identified a core set of industries critical for the country to become self-reliant, along with the science and technology institutions required to sustain these industries.

I will analyse India's vision of self-reliance as set out in its policies in the first few decades after independence, and the factors that led to its successes. I will also analyse some of its limitations and failures. Did we assume that all sectors of technology, with their differing scale and complexity, could be governed with a similar set of policy instruments? Was our approach based on the assumption that technology regimes are stable and slow to change? Did this lead us to create a relatively closed economy, and prove harmful when it was time to show agility, make quick choices? Would we have been better served by planning to become world leaders in specific sectors where we could use our large internal market to secure control over technology and build on this strength? Could these sectors, in turn, have been leveraged to secure technology in other sectors?

Of course, hindsight is always more intelligent than foresight and, to be fair to India's critics, our chosen model of self-reliance carried a whiff of technological autarchy. This was no longer viable in an age when technology was changing rapidly and product complexity was increasing, a new challenge that demanded a retooling of policies. Instead, we decided in the 1980s and 1990s that handing over our public resources to Indian capital and inviting foreign capital to India was all it took to create a growing economy.

The Dynamics of Technology and Self-Reliance   115

Technology was relegated to the status of simply another factor of production that could be bought or sold in the international market. Ergo, the government helps Indian big capital—Ambani, Tata, Birla, Adani—to build big projects, or hands over segments of the Indian market to them, or invites foreign capital to set up plants in India with imported technology. The assumption is that the relocation of plants with advanced technology automatically makes us advanced and 'atmanirbhar' (Hindi for 'self-reliant'). The key difference between an 'atmanirbhar Bharat' and genuine self-reliance is knowledge: who owns the knowledge to build plants and equipment. From chip making to defence equipment, the knowledge of what to manufacture and how is key to self-reliance. Having a factory assemble semi-knock-off kits in India and calling it self-reliance is not good enough (and in the Rafale case, even that does not seem to be materialising).[1] Importing a large plant with all its equipment and designs, whether to produce oil or chips, is not self-reliance, either.

In order to understand technology and its relationship with the economy, we need to understand the inner dynamics of technology and not treat it as just another internal or external factor with a set cost. What differentiates a developed economy from a relatively less developed one is its scientific and technological knowledge. That is why the Netherlands is an advanced country, while Saudi Arabia, with a GDP of a similar order, is not.[2] It is also why the Netherlands

---

[1] Rajat Pandit, 'India Imposes Penalty for Offsets Delay in Rafale Fighter Deal', *The Times of India*, 22 December 2021.
https://timesofindia.indiatimes.com/india/india-imposes-penalty-for-offsets-delay-in-rafale-fighter-deal/articleshow/88420692.cms
Dinakar Peri, 'How Does the Removal of Offset Clause Requirement Affect Rafale-like Deals?' *The Hindu*, 4 October 2020.
https://www.thehindu.com/news/national/the-hindu-explains-how-does-the-removal-of-offset-clause-requirement-affect-rafale-like-deals/article32762645.ece

[2] According to World Bank data for 2021, Netherlands has the 17th highest GDP in the world and Saudi Arabia is 18th.
https://databank.worldbank.org/data/download/GDP.pdf

produces the most advanced chip-making machine in the world,[3] while all that Saudi Arabia exports is hydrocarbons and has to import advanced equipment, plants and processes. Understanding the dynamics of technology can help formulate policies so that developing countries can emerge from their current relations of dependence on more advanced countries: the unequal relationship between recipients and suppliers of technology. It does not follow that any technological development will automatically translate into economic development. Calling technological capability a *necessary condition* for economic development does not imply that it is a sufficient condition as well. The challenge is not only how to develop or advance knowledge, but also how to *integrate it into production*.

The post-World War scenario introduced a number of factors that would change the nature of technological relations between nations. Except for countries like India, which had seen some industrial development, most of the newly independent countries had a very weak technological or industrial base. In this situation, Western nations devised an initial strategy where the developing countries were to supply raw materials and agricultural commodities, while the manufacturing sector would remain largely within the advanced countries. This was termed a *rational division of labour*. It even figured as such in my Class 11 textbook in 1963!

An attempt was made to bring this policy to bear on India. Western nations and manufacturers refused to share with us technologies for steel manufacture, steam turbines, boilers[4] and

[3] Clive Thompson, 'Inside the Machine That Saved Moore's Law', *MIT Technology Review,* 27 October 2021.
https://www.technologyreview.com/2021/10/27/1037118/moores-law-computer-chips/

[4] A comprehensive story of the Indian power sector and self-reliance is yet to be written. As happened with other sectors, the West was willing to supply equipment and set up turnkey projects but not to part with technology. India had asked for this and was turned down by General Electric and Combustion Engineering, both from the US. Associated Electrical Industries

pharmaceuticals,[5] oil exploration,[6] etc. However, the emergence of a powerful socialist bloc willing to transfer technology to developing countries would strengthen their hand, enabling a number of these countries to embark on a path of industrial development with the avowed aim of self-reliance, even as technological expertise remained largely a preserve of the advanced countries. It was when India successfully negotiated with the Soviet Union and East European countries for technology and manufacturing plants that Western companies reluctantly agreed to participate in India's industrial development.

The 'rational division of labour' envisaged by the Western scheme broke down in countries such as India, Brazil, and Korea, but it continued to apply to a number of other Third World countries. Eventually, as some Third World countries broke into the production of manufactured goods, the Western strategy shifted from the supply of manufactured goods to the supply of manufacturing know-how. The West was prepared to part with technology as a set of designs, manufacturing protocols, and equipment required for production, but it would do so under a regime of licenses, joint ventures, or subsidiaries. What was not transferred was the ability to produce new designs and new products. In other words, companies entering into such agreements with Third World entities *retained their monopoly over generating the knowledge required for breaking the dependence of*

  (UK) provided technology for small turbines, which was inadequate for the expansion of India's power program. That is when India approached Czechoslovakia for boilers and the Soviet Union for steam turbines, leading to the formation of Bharat Heavy Electricals Ltd.
[5] The rise of the Indian pharmaceutical sector and the manufacture of antibiotics is described in Section III, Chapter 8, 'The Untold Story of the Left in Indian Science'.
[6] Dipankar Dey, 'State and Foreign Involvement in the Development of Indian Petroleum Industry between 1970–89', Ph.D. Thesis, Calcutta University, 1999 and H.N. Kaul, *K.D. Malaviya and the Evolution of India's Oil Policy*, Allied Publishers Ltd., New Delhi, 1991.
https://ssrn.com/abstract=2317203
http://dx.doi.org/10.2139/ssrn.2317203

their Third World partners. Here was the genesis of a technological dependence that would lead to repetitive collaborations as newer and newer products came into being. The new designs and products were used to keep in line or 'discipline' the Third World partners who lacked the tools to develop them independently.

In response, countries like India raised high protection barriers for the home market, while actively promoting indigenisation. Though much is written today about the failure of this strategy, and much in it can be faulted, there is little doubt that Indian industry gained the ability to produce a variety of goods and develop crucial skills in a whole range of areas thanks to this policy. One has only to look at neighbouring economies and the paucity there of both locally made products and local skills in the entire gamut of industrial goods to realise the significance of India's choice.[7] Even so, India's protected market provided little incentive for its domestic monopolies to advance technology; the gap that opened as a result would prove costly. I have also not dealt here with the more complex role of the Indian big bourgeoisie and the role it played in the development of the technology regime that emerged after independence. Beginning with its willingness to hand over the 'commanding heights of the economy' to the state, it wanted, by the 1980s, a much larger role in the economy.

Amiya Bagchi has perceptively noted that the Indian big bourgeoisie had one foot in industry and the other in trade. How much of this has shaped its attitude towards technology? A protected market means that importing technology is good enough; there is no need then to develop technology. Is this the reason that even when major global players have emerged in India—as have Ambani and Adani—there is little in terms of technology that can be traced back to them? It is not that the Indian big bourgeoise has not developed technology to compete in the international market—Bajaj and Hero are global players in

---

[7] See Ghulam Kibria (1998) on Pakistan, in Section III, Chapter 9, 'Technology in a Postcolonial Setting: Notes from the Subcontinent'.

The Dynamics of Technology and Self-Reliance   119

the two-wheeler market—but such examples are relatively rare. I am not counting the outsourced service providers—software, data, and call centres—as they do not develop new technologies.

While I have gone into some details on the pharmaceutical industry as this relates to the larger theme of this book, of examining technology and the knowledge commons, the question of why the Indian big bourgeoisie did not follow the Japanese or the South Korean big bourgeoisie remains to be addressed. Was it because the Indian market was big enough to create Adani and Ambani to be global players without accessing global markets, unlike the much smaller Japanese or South Korean markets? Was it the one foot in trade and the other in manufacturing that provided the reason for building protected monopolies in India? Or was it that aligning with global capital as a junior partner was incentive enough not to develop technology and compete with the global big boys? I am afraid these, and other such questions have to be left to a future book, and possibly for other authors. This chapter, along with others, is focussed more on technology and therefore has not taken up the complex issues which are, ultimately, determined by the interplay of economics, technology, and politics.

Owing to neoliberal economic policies, the 1990s saw the reversal of self-reliance as an objective, combined with the dismantling of the public sector. The new doctrine was that Indian big capital was capable of participating in the international market for technology on its own, and a larger view of the country developing technology, in terms of knowledge, people, and production capacity, was no longer required. As a party, the BJP, or its earlier avatar of the Jana Sangh, had always opposed planning, calling it a 'socialist disease'. Hence the conversion of the Planning Commission to the NITI Aayog, whose main task is creating papers and presentations on whatever the government decides is the policy of the day, without any long-term vision.

Though the Congress did not officially endorse dismantling the Planning Commission, in practice the 1990s saw government policies dictated by the World Bank far more than the Planning Commission. The policy shift of dismantling self-reliance and the public sector had started under the Congress with Manmohan Singh as finance minister and Narasimha Rao as the prime minister. This accelerated under the Vajpayee government, was taken up again by Manmohan Singh, before the Modi government took a wrecking ball to self-reliance and the Planning Commission.[8] Even while disagreeing with neoliberal policies and the abandonment of self-reliance to develop India's capabilities in science and technology, it must be conceded that there were serious limitations in the way self-reliance was practised, especially the belief that what had worked in the 1950s and 1960s would work just as well today. The increasing pace of technology development carried implications that were not grasped adequately. A policy where the entire supply chain needed to be indigenised was no longer feasible in a more complex world of production. Nor could all the machines required for a plant be indigenised. The challenge was to identify the key technologies in which to build up our capabilities. We needed to move from a global market of unequal exchange to one where we could participate as equals. Understanding the dynamics of technology would enable us to articulate alternate strategies for self-reliance and catch up with the era of rapid technology change, rather than giving up self-reliance itself.

DIMENSIONS OF TECHNOLOGICAL CHALLENGES: SCALE OF PLANTS AND COMPLEXITY OF PRODUCTS

With the new Industrial Policy of 1948, India ushered in a vision of building an independent economy to match its political independence. The need to industrialise and for the state to play a leading role was accepted by almost all political sections; even the Bombay Plan (1944–45) of Indian big capital shared this

---

[8] See Section I, Chapter 1.

The Dynamics of Technology and Self-Reliance   121

vision. The drawing up of the new industrial policy was followed by the founding of the Planning Commission in 1950, the launch of the first Five-Year Plan in 1951, and by the Industrial Policy Resolution of 1956. The final element in this policy structure was the Monopoly and Restrictive Trade Practices Act, 1969. This set in motion two sets of policies. One involved identifying the sectors where the state would play a major role, and what would be left to the private sector. The other was determining what size of plants would be permitted in order to avoid monopolies; in other words, the scale of production permitted to the private sector. Not allowing any private sector company to build a monopoly also involved restricting its production, possibly affecting its economies of scale.

Under these policies, defence, atomic energy, and railways remained exclusively the state's preserve. Only the public sector was allowed into key industries such as coal, iron and steel, shipbuilding, manufacture of telegraph, telephone, mineral oils, etc. While established private sector players were allowed to continue in these sectors, no new players were to enter. Other areas, including textiles, machine tools, chemicals, cement, etc., were left to the private sector while the government retained the right to set certain policy priorities.

The vision behind this policy was that the state alone could mobilise resources on a sufficient scale to rapidly industrialise the country, which was the goal of the independence movement. Heavy industries, energy, steel, communications, were crucial if economic independence was to be realised. Today, we may not recall the extent of underdevelopment in those days. In 1947, India had electricity and telephones in only a few urban areas. Its total generating capacity was less than 1,500 MW, and there were fewer than 80,000 telephones. If this was to change, India needed technology and capital, along with the knowledge to develop them further. Self-reliance was never a static goal to be fulfilled with a one-time import of technology for a particular sector, but a continuous project of developing technology further and matching development elsewhere.

There are many detailed analyses of the successes and failures of the public sector in meeting the twin goals of industrialisation and a self-reliant path of technology.[9] This was not an easy exercise as all decisions on technology imports, partnering foreign companies, etc., were as much about economics and politics as about technology. The second issue is that each of these industries works differently. Setting up steel plants, oil refineries, power plants, etc. is not the same proposition as telecommunication manufacturing: phones, switches, setting up telephone exchanges, and telephone lines.

The scale is a crucial consideration for certain sectors. What would be the optimal size of a power or fertiliser plant? How much would this plant and its size change in future? Should we import plants wholesale as turnkey (ready to operate) plants, or also develop the capability of setting up such plants? If the latter, what equipment would we need to import or manufacture in future? Should we also set up other plants to manufacture the equipment required by these plants?

These issues bring us to Marx's characterisation: he terms as 'Department I' the industry that manufactures the equipment used by 'Department II'; Department II produces goods for consumption. Self-reliance meant that India needed not only steel and power plants to cover the basic needs of its people and industries, it also had to develop industries that would manufacture the machine tools, heavy engineering products, industrial motors, pumps, compressors, etc., required by such plants, e.g., the machinery needed to run power or fertiliser plants. These were key elements in India's goal of self-reliance, particularly after its

---

[9] Amiya Bagchi, 'Public Sector Industry and Quest for Self-Reliance in India', *Economic and Political Weekly*, Vol. 17, 14/16, pp. 615–28, 1982.
https://www.epw.in/journal/1982/14-15-16/annual-number-specials/public-seetor-industry-and-quest-self-reliance-india

The Dynamics of Technology and Self-Reliance

1974 Pokhran nuclear test, when it came under a range of active political sanctions that went beyond the old unwillingness to part with advanced technology.

Instead of looking at these issues in general terms, I am going to take up three sectors to see how far Indian self-reliance succeeded and what its limits were. One of them is the energy sector, of which I have some personal knowledge, having spent a considerable period of my life in this sector. Oil, steel, fertilisers, and chemicals are not identical to power plants, but there are certain similarities among them. They have strong economies of scale and stable technology regimes. Once set up, plants last a long time and are upgraded with incremental changes to existing facilities rather than by scrapping old plants and building new ones. I will briefly touch upon how these sectors faced a similar set of challenges with respect to technology, and how India could have leveraged its huge internal market to become an international player, supplying plant and equipment to other developing countries in these sectors. Instead, it continues to be a recipient of technology; unlike China and South Korea, which leveraged their internal markets to become global players.

In the second instance, I will take up industries where not just scale but the complexity of the product is an element in analysing technology. For example, a modern car has the same purpose as the Ford Model T, the first mass-produced car: both take people from one place to another. The Model T had about 3,000 parts. A car today has about 30,000—an increase by a factor of ten to provide essentially the same function, that of mobility, and using virtually the same technology, the internal combustion engine. Why is this important? While it should be conceivable to mass-produce a car from its raw materials in one massive factory, this is no longer possible with 30,000 components (of which electronic chips constitute 45 per cent of the cost of the car). Here, the complexity of the product introduces new elements when we consider how to look at self-reliance and import substitution. The solution of a near autarchic economy is no longer viable even for continental-sized economies like India and China.

Another example of some relevance is the manufacture of telephone exchanges, particularly in the event of a major advance in technology, such as the shift from analogue technology for switches to digital technology. India had developed the core technology of a digital switch at the Centre for Development of Telematics (C-DoT), and could have become the major global player that Huawei is. Born at almost the same time, C-DoT and Huawei used a similar set of technologies. The difference was state backing, which C-DoT did receive initially, and then did not. Punishing Sam Pitroda for being too close to the Congress party carried the upshot that C-DoT was also punished. Again, it was not technology that failed India's policy, but politics and policy that failed technology and self-reliance.

The third area is of sectors where there is rapid technological change but the *products* are neither complex nor require economies of scale of the kind involved in the process industries. This does not mean they are low-tech products. In fact, some of them come from cutting-edge science, for example, biologic drugs and vaccines. Pharmaceuticals require advanced technology, but they do not need to be produced at the scale of industrial chemicals. The key is technology and, of course, knowledge, or what is called 'intellectual property'. Here, India has performed much better, becoming a global player. Much of this success was due to changes in the Patent Act as also the technological support that India's pharmaceutical industry received from the laboratories of the Council of Scientific and Industrial Research (CSIR). India made major advances in this sector, even though it has regressed lately in bulk drug production—or active pharmaceutical ingredient (API) as it is now called.

The Dynamics of Technology and Self-Reliance

ECONOMIES OF SCALE AND THE RELATIVE STABILITY OF TECHNOLOGY: THE POWER AND FERTILISER SECTORS

If we look at power, fertiliser, or chemical plants, they tend to have stable technologies. Generally, the capital costs are written off over 25–40 years. With nuclear power plants, their life cycle is considered to be 60 years. The average age of coal-fired[10] and nuclear[11] power plants today being 40 years (in the US), replacing existing plants with new ones is not what any country would do. This holds true for oil refineries as well, their average age in the US is again 40 years. Improvements in efficiency would not justify the huge capital costs involved in building new plants. The product of an older plant or a new plant is identical: electricity or a specific chemical. Given their high capital costs, even if the operating costs per unit of output are higher, the cost per unit of output, taking *both* operating and depreciation of capital costs, is lower for older plants. If they have to be modernised, it can be done incrementally through retrofitting. In a sector of this kind, initial collaboration for technology followed by slow indigenisation is a viable route, which was India's self-reliance policy, initially.

It was also understood that domestic entities entering into such collaboration would require building up their knowledge base—knowledge of the processes, design skills, further research and development, and the many intangibles that make up the technology kitty *for the next generation of plants*. It was not enough to just import technology and replicate the existing plants, but important to advance the technologies and therefore produce the

---

[10] International Energy Agency, 'Average Age of Existing Coal Power Plants in Selected Regions in 2020', Last updated 8 October 2021.
https://www.iea.org/data-and-statistics/charts/average-age-of-existing-coal-power-plants-in-selected-regions-in-2020

[11] US Energy Information Administration (EIA), 'What is the Status of the U.S. Nuclear Industry?' *Nuclear Explained: U.S. Nuclear Industry*, 18 April 2022.
https://www.eia.gov/energyexplained/nuclear/us-nuclear-industry.php#:~:text=The%20average%20age%20of%20these,commercial%20operation%20in%20December%201969

next generation of plants. This is why training a cadre of engineers in the power sector, or in the steel sector, was very much a part of India's policy of self-reliance.

A number of young engineers were sent to the Tennessee Valley Authority in the US, to learn how to design and operate power plants. They formed the core of the Damodar Valley Corporation set up in 1948.[12] These engineers would become the backbone of the Indian power sector. Similarly, a team was sent to US Steel and later formed the core technical and management team of the Steel Authority of India (SAIL). Again, when Bharat Heavy Electricals Ltd. (BHEL) was set up, engineers were trained in boiler design in Czechoslovakia and the Soviet Union. Nor was research ignored: BHEL and SAIL had their captive R&D units to improve the processes. One deficiency that persisted was the failure to link up R&D institutions with universities; the academic community became isolated from those who were involved in production. While India did develop processes and technologies in the public sector to expand the size of plants and incorporate other advances, the process was aborted repeatedly. For example, to gain technological self-reliance in the fertiliser or chemical industries, we needed to develop design capabilities, knowledge of catalysts, optimised processes, and equipment-manufacturing capacity. Instead, we imported turnkey plants. Indigenisation took a backseat as the design of plants and processes became dependent on Bechtel, Toyo, and a few other turnkey suppliers. We achieved self-sufficiency in the production of certain segments of fertiliser, such as urea, but at the cost of repeated imports of technology and being unable to become a supplier of plants or technology, whether in India or abroad.

This is unlike what happened in the power sector where equipment manufacture was planned along with the expansion of the sector itself. BHEL received government backing to acquire

---

[12] Daniel Klingensmith, 'O*ne Valley and a Thousand*': *Dams, Nationalism, and Development*, New Delhi: Oxford University Press, 2007.

technology, first from the socialist countries, later from Germany, and the US. By the late 1960s, BHEL had become the leading supplier of power plant equipment in India. This was the route China also followed in building its own power sector. However, unlike India in the 1990s (or since), China did not allow the import of power plant equipment but protected its internal market.

For those unaware of the technology challenges of the Indian power sector and the major contributions that BHEL made, one was adapting the imported technology, whether of Soviet or American corporate origin, to high-ash Indian coals. In the US, 'most steam coals used for electricity require less than 20 per cent ash content (air-dried), and less than 10 per cent ash is required in most utility contracts.'[13] In the US, coal with more than 10 per cent ash content is considered high ash—this level would be the stuff of dreams for India. Indian coals used in power plants have 25 per cent to 45 per cent ash content.[14] Dealing with this much higher ash content required major modifications to the design of boilers and the ash-handling system, which BHEL's boiler design team in Trichy developed—without which India's power program would have stalled. The key issue is that none of the international companies had ever faced the challenge of such high-ash coal; their designs, which India initially used, could not handle the much higher ash content in Indian coals.

Another example is how BHEL developed indigenous technology in control systems. Dr Narla Tata Rao of the Andhra State Electricity Board wanted distributed digital controls—then a new technology—to be implemented in the Vijaywada Thermal Power Station Stage II (now named after him).[15] Starting in

[13] Kentucky Geological Survey, 'Ash Yield in Coal (Proximate Analysis)'. https://www.uky.edu/KGS/coal/coal-analyses-ash-yield.php
[14] Ministry of Coal, 'High Ash Content', *PIB India*, 3 January 2018. https://pib.gov.in/PressReleasePage.aspx?PRID=1515278#:~:text=Ash%20content%20of%20coal%20produced,of%20coal%20deposits%20in%20India
[15] Disclosure: I was a part of the consultancy team on this project since I was

1974, BHEL had a collaboration with Brown Boveri & Cie (BBC, which later became a part of Asea Brown Boveri or ABB).[16] The collaboration was for control systems but did not extend to the critical core boiler controls known as the furnace supervisory and safeguard system (FSSS). BHEL agreed to use the BBC systems for the FSSS, and also modified the cathode ray tube-based operator information system developed by the Electronics Corporation of India Ltd.—thus achieving control as well as indigenising both the integral boiler and turbine controls. This was before either Combustion Engineering or Siemens (their boiler and turbine collaborators, respectively) had done so. To sum up, India used its very large domestic market to secure transfers of technology on favourable terms to BHEL. India's emerging market power helped secure technology that US companies such as Combustion Engineering and General Electric had earlier refused to share. This was a route India would later abandon, after catching the neoliberal fever.

The boiler collaboration with Combustion Engineering (US) and the turbine collaboration with Siemens (Germany) was virtually forced on BHEL, as the National Thermal Power Corporation (NTPC) had World Bank loans that stipulated global tendering and foreign consultants. BHEL realised that certain parameters such as the heat rates and efficiencies specified favoured foreign suppliers and therefore decided to tie up with Combustion and Siemens. I am not analysing here whether the path chosen—collaboration with Siemens or Combustion Engineering—or choosing a more independent path of developing

with DESEIN Pvt. Ltd., one of India's major consultancy organisations in the power sector, and was closely involved with the specifications and subsequent designs of the system for the plant. This was the first example of a distributed digital control system for a power plant in India, and one of the very few in the world at that time.

[16] John Surrey, 'Electric Power Plant in India-A Strategy of Self-Reliance', *Economic and Political Weekly*, Vol. 23, 8, 20 February 1988.
https://www.epw.in/journal/1988/8/special-articles/electric-power-plant-india-strategy-self-reliance.html

boilers and turbines from their existing designs was the right way to go.[17] But having taken this step, BHEL indigenised technology to a level where they could change or modify it to suit their needs, and built a platform on which they were globally competitive. They developed the necessary set of skills not only to indigenise but also develop these technologies further. In contrast, the policies set out for the power sector in the 1990s, of private players allowed free entry with their foreign partners, and even with imported fuel like LNG, damaged not only the power sector but also BHEL's potential to emerge as a leading global supplier.[18]

There is also the other saga of a Siemens collaboration sought to be forced on BHEL during the 1980s. This was fought back both politically and by BHEL engineers and the fledgeling science movement. Ashok Rao is with us today, who was very much a part of that struggle, while the two other stalwarts—A. Gopalakrishnan and K. Vijayachandran are not. P. Ramamurthy raised this issue in Parliament, and his second booklet on this topic—*For Whom the BHEL Tolls: Stop BHEL's Dangerous Truck with Siemens*—was a key element of this struggle. Though the over-arching agreement sought to be steered by V. Krishnamurthy and George Fernandes (secretary and minister of industries, respectively) failed due to this resistance, piecemeal collaborations followed later, aborting major advances in designs that the R&D Department of BHEL headed by A. Gopalakrishnan was doing in various areas.

The only area where BHEL could still advance with its indigenisation program was in the nuclear turbines and condensers, as India was under technology sanctions in these areas. The privatisation of the power sector with the policy of

[17] One of the adverse effects of the Siemens collaboration was that a condenser—a key part of the turbine cycle—developed by BHEL's R&D division was abandoned in favour of a Siemens condenser.
[18] The Enron project was based on imported LNG, a foolish decision since we have relatively little petroleum resources and an abundance of coal. The other LNG-based power plant, NTPC's Kayamkulam became an expensive white elephant as did the Dabhol power project.

independent power producers (IPPs) fragmented the market and led to the large-scale import of turnkey plants with no transfer of technology. The net beneficiaries were the Chinese power plant manufacturers—who had first accessed technology through similar Western sources, using China's large internal market just as BHEL had done for India, and who now supplied a large number of plants to the Indian IPPs. BHEL lost not only a significant part of its home market but also lost out on the possibility of emerging as a major international supplier, a role that China and South Korea went on to assume in Asia and Africa. While India had a huge internal market to leverage in its bid for technology, and though it still figures as a global player in the power plant market, it has fallen behind the Dongfang Electric Corporation (China) and Doosan Heavy Industries and Construction (South Korea). That Doosan, with a much smaller internal market, could transition from being a local to a global player shows the opportunities that India, with its indigenous capacity and much larger market, should have seized, but ended up squandering instead.

Nuclear power needs a separate discussion by itself and I am not going to attempt it here.[19] After the 1974 Pokhran nuclear explosion, stringent sanctions were imposed on the Nuclear Power Corporation and the Bhabha Atomic Research Centre (BARC). India was forced to indigenise the entire technology cycle, not just for the reactors but all other equipment that goes into a nuclear plant, from complex control systems down to lowly gaskets. The setting up of the Electronics Corporation of India Ltd. under the Department of Atomic Energy was in part meant to address the issue of control and instrumentation for the nuclear plants, besides helping India's nascent electronics industry. The Indian nuclear power plants also needed heavy water, an isotope of normal water

---

[19] Prabir Purkayastha, 'With Westinghouse Bankruptcy, the Nuclear Energy Story Nearly Over', *Newsclick*, 7 April 2017.
https://www.newsclick.in/westinghouse-bankruptcy-nuclear-energy-story-nearly-over

The Dynamics of Technology and Self-Reliance

as a moderator to slow down the radio-active decay process in the nuclear fuel of the nuclear reactor. While the Westinghouse or GE reactors used light water (normal water) with graphite moderators, the Canadian CANDU reactors used heavy water as the moderators. We needed to build heavy water production facilities for separating heavy water from normal water, which were similar to the process industries like fertiliser. In fact, India's heavy water plants were initially a part of the fertiliser plants before being separated administratively. Much of the technology and equipment was similar to fertiliser plants, an unexpected technology bonus for both.[20]

One of the untold stories of Indian technology concerns the advances that BARC had to make over a range of technologies—from electronics to metallurgy—and what influence this had on technology development in India. While coal-fired plants and nuclear plants have had a relatively stable technology regime, gas turbines have changed some of these dynamics. They are newer and have shorter life spans. Similarly, direct reduction of iron (DRI) plants have a reduced size and are relatively new compared to blast furnaces, the mainstay of the earlier generation of steel plants. I am not treating new technologies that are emerging in power plants and steel making, as we do not know what their life spans are likely to be. Given that coal-fired plants and blast furnaces are greenhouse gas emitters, gas turbines and DRI plants have as much to do with fuel choice—gas instead of coal—as with technology choices. The larger issue of global warming is dictating a certain choice of technologies, no longer decided solely by efficiency or economics.

I am not going into details on the fertiliser industry in India, which followed a similar path to the power sector. The key in the

[20] For example, stainless steel welding was developed indigenously in Bharat Heavy Plates and Vessels for heat exchanges required for Heavy Water Plants. This is only one example. Many more such examples exist, including developing quality controls for the electronics industry, a spin off from developing ECIL's quality assurance programme.

chemical industry, of which fertilisers are a part, is control over the process technology: the catalysts, the chemical cycles, and the optimisation of the process parameters. If a project is set up as a turnkey plant, the suppliers—whether Bechtel (US) or Toyo Engineering (Japan)—control the technology. Here again, the battle for indigenising technology was fought over two issues. Could the Fertiliser Corporation of India (FCI) be trusted with indigenising the process technology? Second, even if the FCI had this capacity by virtue of the government-owned engineering services provider—Projects and Development India Ltd. (PDIL, initially a division of the FCI and separated from it in 1978)—would it not be better to go for a higher capacity plant? Since PDIL did not have the requisite experience here, did it not make sense to seek a new technology supplier? The capability of PDIL was juxtaposed with the issue of size-of-plant in order to sideline PDIL and chose a turnkey foreign supplier.

India's fertiliser requirement is obviously large given the size of its agriculture. Using the size of its internal market, it would have been a simple matter to negotiate technology transfers in exchange for establishing a set of plants. PDIL, which had already pioneered the development of catalysts and optimised the processes, would have been an obvious partner in any such transfer. India had the capacity to build advanced heat exchangers, reactors, motors, pumps, and compressors, all of which are critical components of process plants. The turnkey model of suppliers means they retain control of the process technology. They guarantee the technology and certain output parameters of the process and its efficiencies. It is risk-free in this sense, but also means virtually no transfer of technology in the sense of knowledge, control or autonomy.

The Planning Commission's Working Group on Fertilisers had written in its chapter on R&D:[21] 'PDIL has done pioneering

[21] Ministry of Chemicals and Fertilizers, Government of India, 'Report of the Working Group on Fertilisers Industry for the Twelfth Plan, R&D and Technical Issues in the Fertiliser industry (2012–13 to 2016–17)', Chapter

research over 35 years in the area of catalyst and has significant achievement in the development of practically the entire range of catalyst relating to ammonia production ... In the catalyst field there are such renowned firms like ICI, BASF, UCIL and Haldor Topsoe and PDIL's work in the catalyst field is at par with the best development in these organisations.' Why, then, did India not build an industry that could become a major player within the country and globally? Why did PDIL not emerge in this role? What should PDIL's relationship with the FCI have been, and with the other big fertiliser industry players in the cooperative sector like IFFCO, and with private players? Did the government have any coordinated plan on how to develop technological self-reliance for India to become a global player in the fertiliser industry?

The answer appears to be that there was no such aim of nurturing technology. At vital moments there was even an active interest in sabotaging efforts in this direction. Sudip Chaudhuri has given an account of what happened at a crucial moment in PDIL's history.[22] India had developed the capacity to produce 900 tons per day (TPD) of ammonia, indigenously. PDIL had mastered the process and designs for such a plant. The parent ministry then argued for World Bank loans, a 1,350 TPD plant size, and used the World Bank's stipulation of previous experience in building plants of this size as an argument against PDIL. Therefore, imported technology became the answer. This plant size and imported technology was thrust on the fertiliser industry as a matter of ideology: to open up India and establish self-reliance as an outmoded goal. This was not an isolated event. There were earlier instances where India's self-reliance was nullified. India already had two organisations that could provide consultancy and process design engineering services to the fertiliser industry,

XIII, pp. 186–210.
[22] Sudip Chaudhuri, 'Public Enterprises and Private Purposes', *Economic and Political Weekly*, Vol. 29, 22, 28 May 1994.
https://www.epw.in/journal/1994/22/special-articles/public-enterprises-and-private-purposes.html

the Project & Design (P&D) division of FCI (later spun off as PDIL), and the FACT Engineering and Design Organisation. However, bypassing these two organisations, Engineers India Ltd (EIL), which had been created for setting up oil refineries, was inducted as a consultant to the fertiliser industry. As the H.N. Sethna Committee said, the Bechtel-supported EIL was thrust on the fertiliser industry essentially to allow the import of turnkey fertiliser plants, bypassing Indian capabilities.[23]

ECONOMIES OF SCALE WITH RELATIVE INSTABILITY OF TECHNOLOGY: THE TELECOM SECTOR

It is not true that economies of scale generate relatively stable technology regimes. Yes, if the capital costs are large, the owners of such plants would be unhappy to find them becoming obsolete. But this has been known to happen when new technologies appeared and the obsolete plants had to be discarded. The most important example of such a fast turnover of technology is chip-making. A foundry with the latest technology—say, a <7 nm semiconductor fabrication plant (commonly called a fab), and a capital cost of up to $20 billion[24]—would retain pole position for a mere three to five years before losing its place at the cutting edge of technology. The same applies to the equipment built around such chips: from cell phones to telecom switches. Electronics is the one

---

[23] Biswajit Dhar, 'Technology Development in Fertiliser Industry', *Economic and Political Weekly*, Vol. 30, p. 2090. Letters to Editor, 26 August 1995.

[24] 'A state-of-the-art semiconductor fab of standard capacity requires roughly $5 billion (for advanced analog fabs) to $20 billion (for advanced logic and memory fabs) of capital expenditure, including land, building, and equipment.' 'Government Incentives and US Competitiveness in Semiconductor Manufacturing' in Antonio Varas, Raj Varadarajan, Jimmy Goodrich, Falan Yinug, 'SIA and BCG Report: Strengthening the Global Semiconductor Supply Chain in an Uncertain Era', p. 18, September 2020. https://web-assets.bcg.com/9d/64/367c63094411b6e9e1407bec0dcc/bcgxsia-strengthening-the-global-semiconductor-value-chain-april-2021.pdf

area where capital costs are large and obsolescence is rapid. In all other areas that have large capital costs—power plants, steel plants, chemical plants, refineries—the technology of production lasts for years. Pharmaceuticals are knowledge intensive, requiring a lot of research, but the capital costs are relatively low. The only outlier is the electronics sector, which is *both knowledge and capital intensive*. The consequence of the rapid turnover of technology in chip-making is that any product based on such chips would also face obsolescence a lot faster than in the energy or the chemical sector. This is why new cell phones are released every 12–18 months, as are laptops and tablets.

One industry affected by such periodic obsolescence is the telecom sector, specifically the telephone exchanges, called switches in telecom language. They are switches as they connect a set of people to one another by 'switching' the connections. Earlier, these were physical connections; today they are switched inside the processors, connecting input and output channels virtually, so they are essentially digital switches. The switches are effectively computers now: microprocessor chip sets that implement in software and hardware the functions of a telephone exchange. With wireless equipment tacked onto such digital switches, they become the backbone of the mobile telecom network.

Before the advent of digital switches, the telephone switches were based on different technologies. Computer-based switches, and then microprocessors, changed the core of telecommunication systems. This also provided a window of opportunity for new players to leapfrog, catching up or moving beyond more established players. In the 1980s, Sam Pitroda's proposal was to develop a small digital 128-port switch, which would serve India's rural areas and could then be scaled up to service up to 10,000 lines. Going beyond 10,000 lines would mean conflict with the agreement on E-10B switch systems that the Industrial Training Institute (ITI) had with Alcatel, the French company which was India's partner in the digital switches. The rural switch was engineered from the beginning for a rugged and non-conditioned environment, unlike the existing switches from major telecom suppliers, which were based on large computers. Pitroda's C-DoT team had to make their switch heat, dust, and monsoon proof, a major technological advance on existing switches. As he writes, 'My plan was to design every component of the small exchanges so that we could incorporate them as modules into the larger exchanges. We'd be using the same cards, processors and software. Everything was modular, flexible, expandable, scalable—and affordable.'[25] C-DoT's target was initially a private automatic branch exchange (PABX) which could also act as a rural switch and be able to withstand exposure to the harsh, ambient conditions of a rural environment, the heat, dust, and monsoon. This was very similar to the route South Korea took, except that ETRI, the South Korean government entity, enjoyed far more state support than C-DoT did. For those interested in how Samsung emerged as one of the major telecom players in the world, a history of telecom in South Korea would show the path followed: how Samsung succeeded with government support and government-generated R&D incorporated in its products, digital switches at the core of it. Interestingly, the Chinese telecom giant Huawei took a very similar path to Samsung, the one that C-DoT had taken in India.

C-DoT succeeded, developing a rural switch that could be scaled up to 10,000 lines (actually 16,000 lines, but it had to be restricted to 10,000 on account of the Alcatel-ITI agreement).[26] Its rural switch was rugged, could work under adverse conditions and won worldwide attention. Though the C-DoT founders have emphasised their own role, the Department of Telecom's policies and the Telecommunications Research Centre were also helpful in inducting C-DoT switches in rural exchanges, giving it a

[25] Sam Pitroda, 'Path to Development, Section 2', *Sam Pitroda Blog,* 8 August 2019.
https://www.sampitroda.com/new-blog-1/path-to-development-section-02
[26] G.B. Meemamsi, *The C-DoT Story: Quest, Inquest, Conquest.* Kedar Publications, New Delhi, 1993.

huge internal market. C-DoT also successfully transferred the manufacturing technology to a number of independent players, while retaining its role as a technology developer, a role that Bell Labs has played in the AT&T set-up. With the success of the C-DoT switches, international companies in wireless business had proposed tie-ups with C-DoT to develop the switching equipment for mobile networks. This was the route Samsung and Huawei followed to become world-leaders in mobile networks, and later also in producing mobile phones or handsets.

I am not going to argue a what-if scenario or say that if only C-DoT had been supported and not made a political casualty of Pitroda's closeness to Rajiv Gandhi, India's telecom story might have been different. Of course, it is also true that without Rajiv Gandhi's support, C-DoT might not have taken off at all. This example is only to show that new windows of opportunity appear with new technologies. India had the same opportunity that China and South Korea did. Why did they succeed but India did not? It is in this context that India's technology policy, its potential and limitations, are important to consider.

The C-DoT story is that a new digital technology in telecom meant a window of opportunity to catch up with more established players in technology. C-DoT managed to scale up from a 128-port switch to a 16,000-port switch and to transition from a small exchange to a large one. It had created a more advanced technology than what was on offer from Alcatel—based on large computers and rapidly becoming obsolete with the advent of microprocessor-based digital switches. C-DoT had the potential to emerge as a Huawei or a Samsung, if only a proper policy framework and support had existed. The problem was not how to develop the necessary technology; C-DoT had succeeded there. Transitioning to mobile networks would have involved further development and raised another set of technical challenges, but the base had been created and the potential existed. This was aborted by India's subsequent policies.

The lesson from the C-DoT experience is that the evolution of technology is not a slow accretion of knowledge and experience. It is marked by breaks, and when such breaks occur, they allow smaller, nimble organisations an entry. Whether it can scale up an initial success with technology to emerge as a major player depends on support from the state and its policy framework before economies of scale can intervene to make it difficult for new players to enter. Using such a window, Samsung and Huawei succeeded in becoming world-class players; C-DoT, with very similar technology, was kneecapped by India's policy makers and did not.

The other advantage India had in securing technology was its huge internal market. China used this advantage and forced telecom switch manufacturers to transfer technology[27] to Chinese companies.[28] India chose private players for telecom services, particularly in mobile services. The private players wanted cheap equipment and handsets to expand their services. Indian vendors of equipment faced reverse protection, favouring foreign players: duties on components were higher than duties on finished products.[29] Not only did India not make a significant attempt to develop equipment manufacture in the country, it did not use its advantage of a very large internal market to bargain for technology either. The privatisation of telecom services, as with private entry in power generation, meant that India did not use

---

[27] Ulaş Emiroğlu, 'Catch-up with Generative State: Lessons from Chinese Telecom Equipment Industry', *Science and Technology Policy Research Centre*, STPS-WP-15/03, Ankara, Turkey.
https://stps.metu.edu.tr/en/system/files/stps_wp_1503.pdf

[28] Robert D. Atkinson, 'Who Lost Lucent?: The Decline of America's Telecom Equipment Industry', *American Affairs*, Vol. 4, 3, Fall 2020.
https://americanaffairsjournal.org/2020/08/who-lost-lucent-the-decline-of-americas-telecom-equipment-industry/

[29] Neha Gupta, Priya Kumar, Rachit Saini, 'India's Electronics Industry: Potential for Domestic Manufacturing and Exports', *ICRIER*, August 2021.
http://icrier.org/pdf/India-Electronics-Industry_2021.pdf

the strategic advantage of its internal market, one of the biggest in the world. Ironically, in both cases, the outcome did not favour the Western countries who had pushed for such policies but ended up benefitting the Chinese equipment and handset companies, with South Korea coming in second!

PHARMACEUTICALS: INDIA BECOMES THE GLOBAL PHARMACY

Unlike other sectors where India squandered its potential, its pharmaceutical sector has become the global pharmacy, supplying generic drugs not only to poor countries but also the rich.[30] Even in vaccine technology, dominated by rich countries and global pharma majors, India's Gennova, a small biotech company based in Pune, has received emergency use authorisation from the Drug Controller General of India for its mRNA vaccine.[31] Developed by this small company, the Gennova vaccine has a big advantage over other mRNA vaccines in that it can be stored at 2–8 degrees Celsius, unlike other mRNA vaccines that need an ultra-cold chain that does not exist in most developing countries. The bulk of the world's vaccines are manufactured in India: Serum Institute is the world's leading vaccine manufacturer by volume. In dollar terms, however, Pfizer is way ahead of the Serum Institute. The difference is that Pfizer holds the patents over the newer vaccines and controls the market of rich countries.

What is the difference between the pharmaceutical sector and other sectors? Why was this Indian industry able to transition

---

[30] Rory Horner, 'The World Needs Pharmaceuticals from China and India to Beat Coronavirus', *The Conversation*, 25 May 2020.
https://theconversation.com/the-world-needs-pharmaceuticals-from-china-and-india-to-beat-coronavirus-138388

[31] Anuradha Mascarenhas, 'Pune-based Gennova Biopharmaceuticals' mRNA Vaccine gets DCGI Nod', *Indian Express*, 29 June 2022.
https://indianexpress.com/article/india/dcgi-indigenous-mrna-covid-jab-emergency-use-covovax-7997142/

Figure 6.1: Manufacturer share by global value (left) and volume (right)

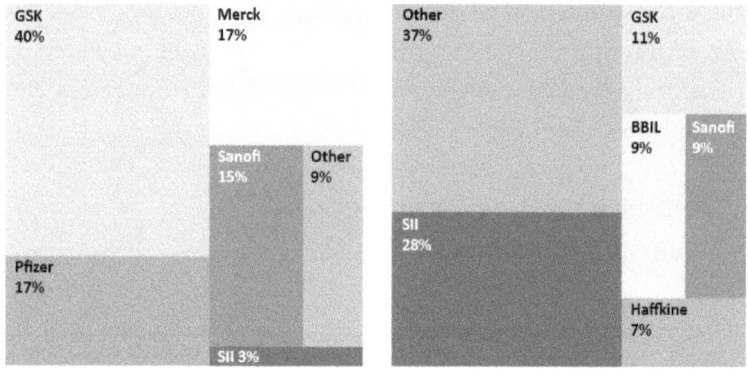

Source: 2020 WHO Global Vaccine Market Report.

successfully from a state of dependence to being the global pharmacy, manufacturing the major part of the world's generic drugs?

The answer, or, to be more correct *answers*, are the following. One, pharmaceutical productions are comparatively smaller in quantity, therefore the scale of production facilities required is much smaller than for example in the chemical industry. A much smaller scale of production means the capital costs of setting up a pilot plant or even the actual production facility are much smaller, reducing the complexity of the system (along with capital costs). Putting together a new production facility is neither prohibitively costly nor does it take too much time. The smaller facilities and costs mean that any entity, whether the state or a private player, can take an experimental risk in setting up new units.

The key issue in pharmaceuticals is the intellectual property rights regime. Ownership of patents and trade secrets are the mechanisms through which the global majors control the industry. Since patents are more difficult to establish with vaccines and biologics, the game over intellectual property rights has moved from patents to asserting bioequivalence standards, with the US

The Dynamics of Technology and Self-Reliance 141

and EU leading the way. As in GATT and later the WTO, the US and the EU have been torch-bearers for their pharma companies.

I have dealt with the intellectual rights issues in pharma in Chapter 8 of this book. A more detailed account is available in a book that two of my colleagues and I put together: *Political Journeys in Health: Essays by and for Amit Sengupta* (2021). Here I only want to note that this battle has been concentrated at the policy level and that of modifying laws, and is about freeing existing knowledge from the control of capital—making it possible for anybody with the requisite knowledge to access the market without property laws on knowledge interfering with this right of access.

There are two issues I want to discuss; both pertain to the size of production and the nature of the product. If we take out vaccines and biologic drugs, technically speaking the pharmaceutical industry is a subset of the chemical industry. This is how it was regarded in earlier times. What made it different is the scale of production and the variety of its output. Even if the processes and equipment are similar, a pharmaceutical compound requires knowledge of each of the specific processes for producing it, as also the equipment necessary for the process. Although a chemical industry, the scale of pharmaceutical production meant that equipment—pumps, pressure vessels, reactors, distillation columns, heat exchangers—could be more easily procured than for producing large quantities of a chemical such as fertiliser. The key for the pharmaceutical industry is the process, and not so much the equipment. The National Chemical Laboratories, Pune, and Indian Drugs & Pharmaceuticals Ltd., Lucknow, provided the process know-how to India's pharmaceutical industry.

The products of the pharmaceutical sector divide into two: small molecules and large molecules. Chemically, a small molecule consists of a compound that does not have too many constituent atoms and is therefore of a low molecular weight; a bigger complex of constituent atoms results in large molecules. In the pharmaceutical sector, some of the new anti-cancer drugs, or monoclonal antibodies against infections, are examples of such large molecules. Biologics define the size of the molecule as the basis of classification, not only with these new drugs, but also older ones like insulin. In terms of their production processes and size of molecule, vaccines are a biologic but are not classified as a drug since they are preventive and not curative—therefore, technically not a medicine.

While biologics might be a new class of drugs for the pharma industry, biologic processes are not different from those used in producing vaccines and antibiotics. The industry already has some experience in terms of the equipment and processes for producing biologic drugs. Again, the key is knowledge of the processes and not the equipment—which is not very different from that used to produce insulin, vaccines, or even penicillin (where biologics processes are used). From the process-engineering point of view, therefore, the older technologies of producing vaccines and insulin are not significantly different from many of those involved in biologic drugs.

The reason the pharmaceutical industry offers easier entry than its allied counterpart, the chemical industry, is the size of the production facilities and the availability of equipment. On both counts—the entry cost and availability of equipment—the pharmaceutical industry provides an easier launch pad for a fast-seconds approach. (The second-mover or fast-second strategy is where a competitor waits for a dominant design to emerge and quickly replicates it with a few value additions.) This is why the bottleneck in pharmaceuticals is not technology or scale of production, unlike sectors such as power and chemicals, but primarily the issue of market access and property rights.

While India's performance as the world's generic drug supplier is welcome news, the catch here is that India is not the leading supplier of what we used to call bulk drugs, now known as active pharmaceutical ingredient (API). In APIs, China is the global leader. This is not because India cannot manufacture the APIs, but that China does it cheaper. Without API capabilities, the Indian drug industry will remain dependent on Chinese supplies. This is not a technological issue but a commercial one.

HOW CAN DEVELOPING COUNTRIES CATCH UP IN TECHNOLOGY?

In an important paper published more than thirty years ago, Carlota Perez and Luc Soete had talked about how countries can catch up in technology using a fast-seconds approach.[32] In an article in 1994 I had discussed how this approach can be used to build self-reliance even in conditions where technological change is much more dynamic.[33] In 2003, I had followed it up with another piece where I argued, using the C-DoT switch as an example, that India could develop technology on par with global corporations. At the time 75 per cent of the installed switches in the country were from C-DoT. Though much has changed since, the conclusions I had offered remain valid. The lesson of C-DoT is that it is possible to compete with the MNCs provided we are prepared to take certain measures. One is to focus on areas where we have technical strength. Obviously, software-based systems are one such area. The second is that there must be a large market we can use for testing and advancing technology in order to become world leaders. The third is to choose areas that offer low entry costs. It is obvious that automation technologies offer such an area, just as telecom does.

The key here is that such technological development will not happen automatically. Those who are arguing for the market as the

---

[32] Carlota Perez and Luc Soete, 'Catching Up in Technology', in Giovanni Dosi, Christopher Freeman, Richard Nelson, Gerald Silverberg, Luc Soete (eds.) *Technical Change and Economic Theory*, Pinter, London, 1988.

[33] Prabir Purkayastha, 'New Technologies and Emerging Structures of Global Dominance', *Economic & Political Weekly*, Vol. 29, 35, 1994.
https://www.epw.in/journal/1994/35/review-industry-and-management-uncategorised/new-technologies-and-emerging

driver of technology do not understand that the market will not create such automation technologies.

If we do not want to de-industrialise, we need public domain science and technology to meet this requirement.

Though much else about the technology scenario has changed since, economies of scale have not disappeared in process industries or in the power sector. Looking back on the last thirty years, it is clear that we need to develop our views on self-reliance and how countries like India or China can catch up in technology and become tomorrow's world leaders, as China has already done in 5G technology. While India and China have their continental-sized internal market, how can countries that have a smaller economy build self-reliance? This is possible if they become part of a larger market, e.g., EU or ASEAN, and become leaders in certain areas of technology. This is what a number of small European countries are doing and what South East Asian countries are hoping to achieve by means of their large regional market.

Perez and Soete identify four phases of growth and argue that, for a developing country, a fast-seconds approach, identifying when a new technology is at its take-off stage, is the most appropriate. It allows for weeding out most unsuccessful technologies and the cost of entry is much lower than entering in the mature phase of technology.

Our discussion of the power, fertiliser, telecommunications, and pharmaceutical sectors has brought the aspect of scale into the picture. Technology development approaches also need to take into account the scale of production and the ability to produce the equipment that goes into a plant. Transfer of technology to run the plant is one part of the requirement, the ability to manufacture such plants—power or fertiliser plants—is also a part of self-reliance. Such a path involves backward linkages, creating the appropriate support structure to guarantee independence, and therefore the state plays an important role in creating conditions for the successful transfer of technology. To put this in Marxist

The Dynamics of Technology and Self-Reliance

Figure 6.2: Four Phases of Technology Development

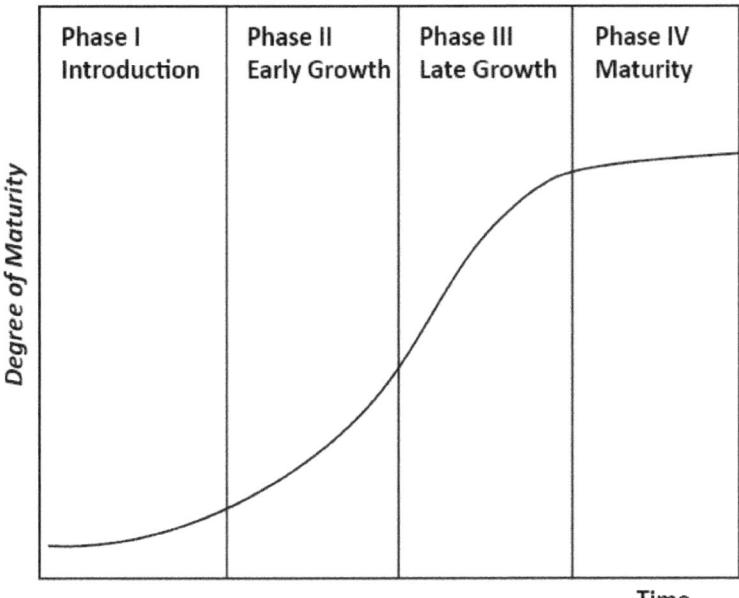

Source: Carlota Perez and Luc Soete, 'Catching Up in Technology', in Giovanni Dosi, Christopher Freeman, Richard Nelson, Gerald Silverberg, Luc Soete (eds.) *Technical Change and Economic Theory*, Pinter, London, 1988, p. 471.

terms, for self-reliance it is not enough to build large process plants in Department II. Equally important is the capability to produce the associated equipment: Department I, that produces the equipment for process plants in Department II.

If these conditions are met, such a path of self-reliance becomes possible. It needs the support of the state. The success of India's self-reliance was in building the chain of equipment production, from plants that built the equipment to the process and power plants that delivered products and electricity, even if this success arrived slower than anticipated.

The weakness was that the links between research in academia and industry remained weak. Technology was understood as simply the import of plants and the ability to operate them. Reproducing such plants, modifying them and advancing their designs, both in terms of size and higher efficiencies, was not recognised since technology was assumed to be static rather than a continuous process. The relative absence of knowledge creation that linked up with industry, i.e., the integration of new knowledge with the production process, became the weak link in our effort towards self-reliance.

While India did use the size of its market to leverage technology till the 1970s, this was weakened in the 1980s and given up with (neo)liberalisation in the 1990s. By embracing neoliberalism, India effectively abandoned this path of development. In contrast, China used its market to develop its capacity for producing power plant equipment and eventually became the world's leading supplier. South Korea, with a smaller market, has also emerged as a major global supplier. India, which had a head start over both with BHEL, lags behind them today. The lesson here is that given the size of their internal market, it is possible for countries to bargain for technology and to enter even at (Perez and Soete's) Phase IV, or the mature phase of technology.

In telecom, where a major technology change took place, a fast-seconds approach—or Phase II in Perez and Soete's formulation—was very much on the cards. This is what Huawei did in China and Samsung in South Korea. Given the size of the global market and the rapidity of change, wireless or mobile technology moving from 1G to 5G in a scant two decades, India missed the boat completely. This in spite of having set up C-DoT and its early success in building microprocessor-based telecom switches before Huawei and Samsung.

The only success story India has in hand is its pharmaceutical sector, where the size of the plants is small, with lower entry-level costs, and also the fact that the 1970 Patent Act endured till 2004 and allowed India's generic industry to become the world's major supplier. Here, the scale of the plants and the lower entry costs helped a large number of private players to emerge. The AIDS pandemic brought to the fore the connection between public health and private profits in the pharma sector, which is being driven home again during the COVID-19 pandemic.

In conclusion, technology is not static. Its development takes place not just through continuous evolution but also significant breaks. In addition, size and complexity are important dimensions in the development of technology. It is always possible to enter an industry either through the import of technology or a fast-seconds approach. The requirement is to understand the dimension of the technology we are trying to develop, the industries required to sustain it, and the knowledge required for its future development; a one-time development or a one-off import of technology will not suffice. Finally, technology is knowledge, and knowledge requires people as much as it does institutions and capital.

Section III

Mapping Public-Interest Science and Technology

7. Science in the Light of Social History

Do scientific laws 'reflect' the objective material reality outside us or are they built subjectively by us, to theorise our observations and experiments? For instance, Newtonian mechanics tells of how the solar system and planets are bound together by the forces of gravity and their own momentum. We can argue that force, gravity, and momentum are concepts we have created; at the same time, they express relationships in nature—such as the solar system—that do not simply make up a subjective picture we have created. They exist independently, outside of us.

While accepting that science expresses *real relationships* in nature, Marx, and most Marxists, also hold that science and scientific knowledge are *historically* produced. The developments in science are not just the works of a few great men who, with the power of their intellect, laid bare the secrets of nature. Science is created historically: the *needs of society* give rise to science, technology and the instruments of enquiry used in scientific experiments, including—as in astronomy—the observation of nature.

Marx noted of his fellow Young Hegelian, Ludwig Feuerbach, that he 'speaks in particular of the perception of natural science; he mentions secrets which are disclosed only to the eye of the physicist and chemist: but where would natural science be without industry and commerce? Even this "pure" natural science is provided with an aim, as with its material, only through trade and industry, through the sensuous activity of men'.[1] And Engels adds, 'If, as you say, technique largely depends on the state of science, science depends far more still on the state and the requirements

---

[1] Karl Marx and Frederick Engels, *The German Ideology*, 1845.

of technique. If society has a technical need, that helps science forward more than ten universities.'[2] Pointing to how the needs of society drive science, Engels was careful to talk also about levels in science—he discussed how it moved from astronomy to mechanics, then to chemical and finally the biological sciences. In *Science in History*, J.D. Bernal recognises the inner complexity of science, with chemistry and biology adding newer dimensions.[3] He was quite aware that discoveries in genetics, for example, would not take place without first developing physics and then chemistry. In the Marxist scheme, the external needs of society drive science, but science does not drive discovery as it pleases. There is an *inner logic of discovery as well*, which determines what set of problems are solvable within the science and technology of a given age.

SCIENCE AS A PRODUCT OF HISTORY: THE PRODUCTION OF IDEAS

It is hard to understand the impact of Marxist thought on the history of science—viewing science in a socio-historical context—without going back to 1931. In London that year, at the Second International Congress of the History of Science and Technology, a team of brilliant Soviet scholars led by Nikolai Bukharin presented the view that science is a product of history.[4] This had enormous impact. For the first time, the history of science was presented as a social and historical process rather than an aggregate of individual triumphs. For a band of young scientists, including

---

[2] Engels gives the specific example of hydrostatics developing from the need to control mountain streams in Italy, the use of electricity driving scientific knowledge about it. Letter from Engels to Borgius, 25 January 1894.
http://www.marxists.org/archive/marx/works/1894/letters/94_01_25.htm

[3] John D. Bernal, *Science in History*, Vols. 1 to 4, London: Watts & Co, Reprinted by KSSP, 1954.

[4] Nikolai I. Bukharin et al, *Science at the Cross Roads: Papers Presented to the International Congress of the History of Science and Technology, held in London from 29 June to 3 July 1931*, 2014. [Originally, Joseph Needham (Preface), P.G. Werskey (Introduction), Frank Cass and Co, 1971.]

J.D. Bernal, Joseph Needham, J.B.S. Haldane, and Lancelot Hogben, this was a completely new outlook. In a seminal paper presented at the Congress—'The Social and Economic Roots of Newton's *Principia*'—the Soviet physicist Boris Hessen applied 'the method of dialectical materialism and the conception of this historical process which Marx created' to analyse 'the genesis and development of Newton's work within the context of the period in which he lived'.

Making science a socio-historical process raised the question of whether science could be characterised by the class interest it served, as bourgeois science or proletarian science. (The benefit of hindsight has given us greater clarity on this and other related questions.) How did class society govern the categories of thought in which scientific laws were cast? In other words, isn't ideology inevitably a part of science, producing a class-based science?[5] Are the laws of science reflections of an objective world? Or are they simply social constructs reflecting class forces in society? Much of this debate came about with the theory of relativity, quantum mechanics, and, later, with Trofim Lysenko's infamous opposition to genetics in the Soviet Union.

A very important section of scientists, particularly the Copenhagen school, used the results of the theory of relativity and quantum mechanics to deny objectivity to the external world.[6] They considered the laws of science as laws of experiments created by scientists. Whether these laws had an objective basis was, for them, not a valid question. The issue of materialism and the fight against idealism in science were very much part of the ideological

---

[5] The social constructivist school of thought denies (or is at best agnostic on) the objective existence of laws of nature, regarding them as a shared belief system of a group of people, in this case of the scientific community. W. Detel, 'Social Constructivism', in *International Encyclopedia of the Social & Behavioral Sciences*, 2001.
https://www.sciencedirect.com/topics/psychology/social-constructivism

[6] Lenin wrote about this crisis in physics in his *Materialism and Empirio-Criticism*, 1909, and later in his *Philosophical Notebooks*, 1913.

struggle being waged among scientists at the time. For a certain section of Soviet philosophers of science, there were two ways of looking at science—the bourgeois way and the proletarian way. An excellent account of these issues is to be found in Loren Graham's *Science, Philosophy and Human Behaviour in the Soviet Union*.[7] As Graham wrote, he is virtually the only scholar in the West to have examined the philosophical issues addressed by scientists and philosophers in the Soviet Union, without reducing the whole of Soviet philosophy of science to a single debate about Lysenko. What was missed was that the *interpretation* of scientific laws according to the ideological beliefs of scientists is quite distinct from the argument that scientific theories must conform to a political ideology. It was the interpretation that needed to be changed, not the laws of nature.

Major advances in science always challenge existing philosophical systems, an effect not restricted to the Soviet Union. One of the reasons Einstein's theory of relativity was denied the Nobel Prize was that the major philosophers disagreed with his formulation that space and time are intertwined. Debating the theory of relativity at the *Société française de philosophie* in April 1922, Henri Bergson, a leading light of philosophy in the first half of the twentieth century, argued that Einstein was philosophically wrong and it was in philosophical terms that time had to be understood. Einstein responded that there were only two valid ways of understanding time: physical and psychological. Time as posited by the philosophers did not exist. The science historian Jimena Canales writes:

> When the Nobel Prize was awarded to Einstein a few months later, it was not given for the theory that had made the physicist famous: relativity. Instead, it was given 'for his discovery of the law of the photoelectric effect'—an area of science that hardly

[7] Loren R. Graham, *Science, Philosophy and Human Behaviour in the Soviet Union*, Columbia University Press, 1987.

jolted the public's imagination to the degree that relativity did. The reasons behind the decision to focus on work other than relativity were directly traced to what Bergson said that day in Paris.[8]

Bergson's objection was not the only reason for the Nobel's Committee's decision. Einstein was also disliked by a set of establishment scientists, including German scientists, who hated him for being a Jew and for his pacifism during the First World War. Philipp Lenard, a German scientist who had received the Nobel Prize in 1905, campaigned against Einstein, saying that good science, 'like everything else man produces', was grounded in bloodlines.[9] We still hear much about Soviet science being overridden by ideology and the case of Lysenkoism. Such hazards continue to plague science in the 'West'—take for example the inequities of race and gender that are increasingly coming to light—but they are seldom acknowledged as ideology trumping science.

It is only now that we have begun to address race and gender discrimination in the sciences (in India we must add caste to this brew), and how generations of scholars suffered on this account. They include Rosalind Franklin, whose key discovery led to deciphering the structure of DNA as a double helix but whose contribution was a long time being acknowledged.[10] Or Alice Ball,

---

[8] Jimena Canales, 'This Philosopher Helped Ensure There Was No Nobel for Relativity: Henri Bergson's Debate with Albert Einstein Reached and Swayed the 1921 Nobel Committee', *Nautilus*, 18 April 2016.
https://nautil.us/this-philosopher-helped-ensure-there-was-no-nobel-for-relativity-4554/

[9] Lenard became the chief of Aryan physics under Hitler. Richard Gunderman, 'When Science Gets Ugly—The Story of Philipp Lenard and Albert Einstein' *The Conversation*, 16 June 2015.
https://theconversation.com/when-science-gets-ugly-the-story-of-philipp-lenard-and-albert-einstein-43165

[10] Anne Sayre, *Rosalind Franklin and DNA*, W.W. Norton & Co., New York, 1975.

the African-American woman and chemist who developed the first injectable leprosy drug but was denied recognition.[11] Stephen Jay Gould in his *Mismeasure of Man* gives examples of how the scientists' racist understanding of humans led to systematic biases even in mundane matters like measurement of cranial size![12] In our times, the race debate was resurrected as an IQ debate which lingers to this day.[13]

Much of this debate on ideology and science came to a head in the Soviet Union with the Lysenko issue. While Lysenko's conclusions were bad science—and we need to register this—there was considerable argument within science (which continues even today as the IQ Debate) about the effect of heredity and environment on the individual. The infant science of genetics was predisposed to regarding heredity as more significant than environment; this buttressed the ideological claim that social hierarchy is due to the inherited superiority of certain groups of people. Also, studies in genetics at the time were confined almost exclusively to the fruit fly and did not appear to have any immediate utility for increasing agricultural output—the major challenge before Soviet agriculture. It was in this context that Lysenko, an excellent plant breeder supported by the Communist Party, attacked genetics and set Soviet science back by decades in this sphere. The great advances

[11] Alice Ball distilled the active ingredient from Chaulmoogra Oil, the only treatment known in Asia (particularly India) for leprosy. In Jeannette Brown, *African American Women Chemists*, Oxford University Press, New York, 2012.

[12] Stephen Jay Gould, *The Mismeasure of Man*, W. W. Norton & Company, 1981.

[13] A well-known controversy is around the book by Richard J. Herrnstein and Charles Murray, *The Bell Curve: Intelligence and Class Structure in American Life*, 1994. Detailed critiques are available in Russell Jacoby and Naomi Glauberman (eds.), *The Bell Curve Debate: History, Documents, Opinions*, 1995, and Bob Herbert, 'In America, Throwing a Curve', *The New York Times*, 26 October 1994. Herbert made the most devastating critique. His review called *The Bell Curve* 'a scabrous piece of racial pornography masquerading as serious scholarship.'
https://www.nytimes.com/1994/10/26/opinion/in-america-throwing-a-curve.html

in agriculture with the green revolution of the 1960s could not have taken place without an understanding of genetics.

This indicates the peril of leading science in a particular direction based on judgements of 'bourgeois science' and 'proletarian science', or treating science as an immediate source of increasing production. In the Lysenko case, the urgent needs of Soviet agriculture overrode good science and caused lasting harm to both Soviet agriculture and science.

The social needs that drive science may very well be those of the ruling classes. This is truer today when the relationship between science and technology is much closer. Whether in genetics or microelectronics, advances in knowledge are incorporated as technology much faster than they used to be. When we talk of societal needs driving science, this by no means implies society as a whole. It simply means that the impulse driving science comes from outside science and within society, be it ruling class interests or larger social concerns.

If societal demands drive science, and it fulfils class needs as a part of the forces of production (or destruction, for military purposes), why should we not call it class science? How are we to reconcile these two views—that science is a product of history and yet creates objective knowledge?[14] Finally, is science also a part of the ideology of the ruling classes? These are the questions we must address. While the relationship between science and ideology is simpler to study in the realm of physics and chemistry, it becomes more complex in areas such as psychology, where the social and the biological intersect, allowing far more play to ideology.

If science is historically produced, how does this historical process generate objective science? We need to recognise that while the problems to be solved are posed by social needs, the

---

[14] Richard Levins, 'Class Science and Scientific Truth', keynote address at *The New York Marxist School's Conference on Dialectical Materialism*, held in October 1979. He has a slightly different way of framing the same issue in his keynote address.
https://libcom.org/article/class-science-and-scientific-truth

knowledge thus created is independent of the social needs that led to its development. When ancient agriculturists—Egyptian, Indian, Babylonian, Chinese—needed to know when to sow and when to reap, this created the demand for a calendar. The demand for a calendar translated, in turn, to a study of the heavens and gave birth to astronomy. However, the resulting astronomical knowledge was based on the objective reality of the motions of heavenly objects, and stood independent of the social needs that gave rise to it.

In the natural sciences, for all that social needs do propel science, the science that is discovered expresses a relationship between parts of nature, a relationship encapsulated by the laws of science. Even if stated as abstractions using terms such as mass, momentum, and force, they express invariant relationships within nature that we capture in our thought process. These are what we call the laws of science.[15] Since the laws of science change continuously—with, for instance, Newtonian science replaced by the Einsteinian framework—how do we consider these laws as invariant? It is possible to derive Newtonian science from the Einsteinian one, imposing certain restrictive conditions such as velocities at which objects are moving, etc. Newtonian science is not invalid because of Einstein. In fact, we still use Newtonian formulations in our everyday work when they will suffice for the kind of calculations we are making. Newtonian physics is not false science, only an incomplete one in this sense. For the purpose it was created, it was sufficiently complete. A partial view of nature is inherent in any science, as science is never complete—nature is inexhaustible. Its incompleteness does not make science wrong or invalid, except in the sense of trivial errors.

[15] There are many examples in science where a different set of concepts has led to the same results, for example, Newton's Third Law and Conservation of Momentum. Both lead to the same set of results, provoking the question of whether one is more 'fundamental' than the other. What we are dealing with here is that an alternate basis to known laws can be found via a different set of fundamental properties as defined by us. The results are the same, and invariant, under both frameworks.

The second set of issues arises out of the categories by which science expresses itself. Since these are also historically produced, is ideology not inherently a part of science? The categories of thought that we use to capture scientific phenomena do owe their origin to the human being's social consciousness. But depending on the subject, these conceptual categories—whether mass, momentum, or energy—may have little to do with ideology and more with how these categories have emerged historically.

The picture changes when we come to the science of human beings, where our categories of thought are deeply imprinted with social consciousness.[16] That is why 'scientific' racism and eugenics used science to justify racial exclusion, slavery, and genocide. Even Darwin was a creature of his times, justifying scientific racism while opposing slavery. Agustín Fuentes writes, 'Darwin thought he was relying on data, objectivity, and scientific thinking in describing human evolutionary outcomes. But for much of the book, he was not. *Descent*, like so many of the scientific tomes of Darwin's day, offers a racist and sexist view of humanity'. As did the famous American jurist, Oliver Wendell Holmes Jr., who justified the forced sterilisation of women on the basis of perceived inherited 'imbecility' and eugenics.[17] His chilling words in the judgement (*Buck vs. Bell*, 1927), 'Three generations of imbeciles are enough', were supposed to have damned the victim. They have instead damned him to posterity. Neither is science itself—for example, cognitive psychology—free of ideological bias. But even

---

[16] Marx wrote to Frederick Engels in 1862, 'It is remarkable how Darwin rediscovers, among the beasts and plants, the society of England with its division of labour, competition, opening up of new markets, "inventions" and Malthusian "struggle for existence". It is Hobbes' *bellum omnium contra omnes* and is reminiscent of Hegel's *Phenomenology*, in which civil society figures as an "intellectual animal kingdom", whereas, in Darwin, the animal kingdom figures as civil society.'
https://marxists.architexturez.net/archive/marx/works/1862/letters/62_06_18.htm

[17] Edwin Black, *War Against the Weak: Eugenics and America's Campaign to Create a Master Race*, (Expanded Edition) Dialog Press, 2012.

here, it is important to realise that disputes concerning science and ideology, though mixed intimately, need to be fought *within* science while also exposing its complicity with the social structure of race, class, caste and gender. We cannot reject the discipline because ideological biases exist within the cognitive categories of its formulations, but we must fight as Stephen Jay Gould and Richard Lewontin did in the battle with their Harvard colleague E.O. Wilson, over socio-biology. Recent evidence from personal letters that Wilson exchanged with known proponents of 'race science' or 'scientific racism', bears out the racist roots of sociobiology.[18] Without the critique of Gould and Lewontin, articulated within science, a simple dismissal of Wilson as a racist would not have sufficed. The battle is, of course, not just within science; it is also a battle in society. As long as the historical conditions exist for class rule, science, as a part of this ideological battle, will also be a terrain of struggle. The battle over characterising race, caste, and gender differences as biological or ideological will continue as long as the social basis of these divisions remains in existence.

How does the ruling class impose its views on the development of science?

First, it 'selects' by way of funding or patronage the class of problems scientists should work on. Funding determines which part of science will attract the largest number of working scientists. Predictably, the ruling class chooses the problems of science that are directly relevant to capital or to imperialist domination. The US government-led Manhattan Project (1942–45) made it

---

[18] One such celebrated fight is the text of E.O. Wilson's book *Sociobiology: The New Synthesis*, which was criticised by his Harvard colleagues Stephen Jay Gould, Richard Lewontin, and others for its racist framing. A recent article gives damning evidence of Wilson's racist views from private letters he exchanged with a known figure in racist science, the Canadian psychologist J. Philippe Rushton. Stacy Farina and Matthew Gibbons, '"The Last Refuge of Scoundrels": New Evidence of E. O. Wilson's Intimacy with Scientific Racism', *Science for the People*, 1 February 2022. https://magazine.scienceforthepeople.org/online/the-last-refuge-of-scoundrels/

clear that this is the era of Big Science which abundantly repays state investment by producing weapons of mass destruction. The relationship between the state and the military-industrial complex as a major, if not *the* major driver of research in science and technology, was a product of the Second World War and the emergence of Big Science funded by the state.

Second, the kind of science practised, and its results, are used directly for the increased domination of capital, whether over other classes or natural resources. Science as a factor of production helps to make capital even more powerful. Not only does it provide the means to develop new technology, increasing the exploitation of labour, but also becomes the means of control over the working class or peasantry. Monsanto's Bt seeds, introduced into the market as hybrid seeds, are an example. Control over the seeds gives Monsanto control over the farmers.[19] Bt is *Bacillus thuringiensis*, a soil bacterium that gives cotton plants resistance against a class of pests.

The problem with this mode of pest control is that pests quickly develop resistance—so that the farmer requires newer and newer varieties of Bt seeds. While this problem was identified quite early on, the choice of Bt to be implanted though genetic modifications into various seeds, and as a hybrid, is still governed by Monsanto's corporate needs, not those of the farmer or the consumer. Third, in a range of disciplines, science and ideology are intimately mixed, making any separation difficult. This is especially true in areas where science and truth directly endanger class rule. An example would be, of course, racism. Darwin's theories were used to justify the supremacy of the white race, and, by extension, European imperial rule. Later, IQ studies were used for the same purpose.[20] Social sciences, biological sciences and psychology are particularly

---

[19] Satyajit Rath and Prabir Purkayastha, 'Bt Brinjal: Need to Refocus the Debate', *Economic and Political Weekly*, Vol. 45, 20, 2010. https://www.epw.in/journal/2010/20/perspectives/bt-brinjal-need-refocus-debate.html

[20] The best known is *The Bell Curve*, cited in Footnote 13.

vulnerable to such misuse of the science-ideology mix.

The contradiction—a science governed by class needs, yet discovering real truths about the world—cannot be resolved at the level of ideas. 'The resolution of the contradiction between science as the growth of human knowledge and science as ideology of oppression comes only with political revolution.'[21] It is by resolving the social contradictions within which science develops that we can hope to resolve the contradiction between ideology and science.

SCIENCE, TECHNOLOGY AND INDUSTRIAL PRODUCTION

How does the layman view production? The tendency is to view science and technology as identical, or joined at the hip like Siamese twins. It is important to separate the artefacts—the tools or machines—that are a physical part of the means of production, from the knowledge that enables the creation of these tools. The objective of technological activity is the production of artefacts; scientific activity aims to produce knowledge. Here again, scientific knowledge—the laws of nature—comprises a part of the knowledge required to create artefacts, machines or tools. However, there is a component of knowledge that is technical, not merely grounded in the laws of nature. It may incorporate empirical knowledge without fully knowing why such a relationship exists in nature. Such knowledge can take the form of rules of thumb, or may derive from empirical relationships, the factor of safety, past experience of technologists, or technology laws/formulae/charts, etc.

In Marx's work, as in much contemporary writing, scientific and technological knowledge is loosely clubbed together. Why do we need to differentiate these two forms of knowledge? Scientific knowledge, once created, is neither local nor social. Whatever the place or time, the law of gravitation works in the same way. It does

---

[21] Levins, 'Class Science and Scientific Truth'.

not depend on the availability or cost of materials, an issue that all technology needs to address; nor is it affected by the absence of scientific knowledge. The technical knowledge embedded in an artefact can express simplified relations derived from experiments, using the factor of safety (based on experience) to bridge the gap between our ignorance and what we reason should be safe to try out.

Technology also has to take into account the economics of production, such as the cost of materials, what is available locally, and the workers' pre-existing knowledge of similar artefacts. A simple, perhaps simplistic, example of how cost and not just science determines our choices would be that of electrical cables. We do not make electrical cables out of gold but use copper and lead even though gold is a better conductor. The cost of gold is a part of technological knowledge, a factor in the larger system of production that technology uses to build artefacts. Cost calculations—cost-benefit analyses—are intrinsic to technology in a way they are not to science.[22] The relation between science and production is mediated through technology.

Marx's analysis of capital illustrates his deep knowledge of the history of technology, particularly in the development of capitalism. Marx looked upon technology as the major force in capitalist production.[23] But he was also clear that capitalism

---

[22] Costs of artefacts for scientific experiments/explorations—the Hadron Collider, the Webb telescope—are of course important in science, but the artefacts used in science express the relationship between science and technology, not the relationships within nature that science explores. Once a law/relationship in nature is discovered, it can be used again and again without any loss. Unlike artefacts, laws of nature do not wear out with use. We also do not make a cost-benefit analysis of the cost of the Hadron Collider, which cost $9 billion, versus the benefit of confirming the existence of the Higgs boson.

[23] Nathan Rosenberg, 'Karl Marx on the Economic Role of Science', *The Journal of Political Economy*, Vol. 82, 4, pp. 713-728, 1974. Rosenberg's account of Marx's understanding of science, technology and society is quite a Marxian one, even though he would not identify himself as a Marxist. https://www.jstor.org/stable/1837142

first created the manufacturing stage before moving into *machinofacture*. Once labour processes are broken down by capital into their component processes, technology can create machines to take over the functions of skilled craftsmen. Marx went a step further, to clarify that capital not only creates machines to speed up and intensify the labour process, but also creates machines that, in turn, manufacture machines. And it is technology and machines that bring about the fundamental change in capitalism—its continuous dynamism. With reference to this dynamism, Marx and Engels write in *The Communist Manifesto* (1848):

> The bourgeoisie, during its rule of scarce one hundred years, has created more massive and more colossal productive forces than have all preceding generations together. Subjection of Nature's forces to man, machinery, application of chemistry to industry and agriculture, steam-navigation, railways, electric telegraphs, clearing of whole continents for cultivation, canalisation of rivers, whole populations conjured out of the ground—what earlier century had even a presentiment that such productive forces slumbered in the lap of social labour?

In *A World to Win: Essays on The Communist Manifesto*, Aijaz Ahmad writes:

> Today's reader tends to think not of the capitalism and colonialism of Marx's time but the capitalism and imperialism of our own time . . . . Marx speaks here of capitalism's drive to unify the globe through a 'revolution' in transport—at a time when steamboats and railways were a bare novelty. The first steamboat had sailed from the Americas to Europe in 1819, and as late as 1840 railways in England itself covered merely 843 miles of track.[24]

[24] Aijaz Ahmad, 'The Communist Manifesto: In Its Own Time and In Ours', in Prakash Karat (ed.) *A World to Win: Essays on The Communist Manifesto*,

Science in the Light of Social History

It is difficult to conceive how Marx, living in societies that were still overwhelmingly rural, could yet imagine the modern world created by capital, science, and technology. Regarding the creation of this technology of building machines, Marx also clarified that such a change could not have arisen by experience and rule of thumb alone. Scientific knowledge was indispensable in this process. Marx says, 'The implements of labour, in the form of machinery, necessitate the substitution of natural forces for human force, and the conscious application of science, instead of rule of thumb'.[25] Also: 'Intelligence in production expands in one direction because it vanishes in many others. What is lost by the detail labourers, is concentrated in the capital that employs them [. .] modern industry makes science a productive force and presses it into the service of capital'.[26]

Marx also talked of science as universal labour. In his scheme, capital converts scientists to wage labour, just as it does everyone else, from physicians to artists and poets.[27] Since science is one product that does not change based on either local or social conditions, the labour that produces it—scientific labour—is in this sense closest to the concept of universal labour Marx used in his labour theory of value. He also distinguished this scientific, universal labour from cooperative labour.

LeftWord Books, 1999.

[25] Karl Marx, 'Division of Labour and Manufacture' in *Capital* Vol. I, 1867. http://www.marxists.org/archive/marx/works/1867-c1/ch14.htm and Karl Marx, 'Machinery and Modern Industry' in *Capital* Vol. I, 1867. http://www.marxists.org/archive/marx/works/1867-c1/ch15.htm

[26] I have taken the preceding two quotes from Robert S. Cohen, 'Karl Marx on Science and Nature (Excerpts)', *Science as Social Process*, Workshop Discussion, Washington DC, 1978. http://www.autodidactproject.org/other/sn-cohenrs1.html

[27] 'The bourgeoisie has stripped of its halo every occupation hitherto honoured and looked up to with reverent awe. It has converted the physician, the lawyer, the priest, the poet and the scientist into wage labourers'. Karl Marx and Frederick Engels, 'Bourgeois and Proletarians' in *The Communist Manifesto*, 1848.

Universal labour is scientific labour, such as discoveries and inventions. This labour is conditioned on the cooperation of living fellow-beings and on the labours of those who have gone before. Cooperative labour, on the other hand, is a direct co-operation of living individuals. (*Capital*, Vol. 3)

In Marx's formulation, scientific and technical knowledge were both intellectual labour and so a part of universal labour. Without drawing a detailed relation between scientific and technical knowledge, suppose we club together both forms of knowledge in this historical account of the development of productive forces? One could argue that Marx has over-emphasised the scientific aspects of building the technology of machines; that he is talking about the largely craft-based knowledge that created the first set of machines. This, however, is a matter of detail.

One of the major points in Marx's treatment of science and technology is that he recognises that capital, in its search for profits, is continuously revolutionising the means of production: change in the means of production is built into the capitalist system. This is unlike any other ruling class, where the stability of the production system was the goal of the ruling classes.[28] The Indian caste system is a particular example of this stability, where no new technology could be admitted as people were tied to their occupations based on a minute division of labour. The only way to introduce new technology into the system was to change one's religion. It is not an accident that the new technologies that arrived in India with Islam saw large-scale conversion to Islam by artisans in order to practise those crafts.

The developments of machinofacture, particularly in textiles in England, meant that there was an enormous development in

---

[28] 'The bourgeoisie cannot exist without constantly revolutionising the instruments of production, and thereby the relations of production, and with them the whole relations of society. Conservation of the old modes of production in unaltered form, was, on the contrary, the first condition of existence for all earlier industrial classes'. Marx and Engels, ibid.

the science of mechanics. Not only did the scientists have more things to study, the technology that was developing also provided new instruments of discovery. The chemical industry in France developed soon after, along with the dye industry. They provided further impetus to scientific and technological research.

The scientific revolution of the sixteenth to seventeenth centuries was the product of the renaissance, which had brought together a number of elements: a combination of craft or artisanal knowledge with the development of mathematics. New instruments were developed, such as the telescope and the magnifying glass, bringing new phenomena into view. Finally, access to knowledge increased enormously with the printing press. All this engendered the flowering of art/culture and science. But let it not blind us to the brutality of capital, its genocidal march through Africa and the Americas, the loot and mass enslavement of people. Without this loot, slavery, and genocide, there would not have been the post-renaissance Europe we know, or the industrial revolution.

The technological revolution followed the scientific revolution. If the scientific revolution dates to the sixteenth and seventeenth centuries, and was coterminous with the development of capitalist relations in production, the technological revolution—machinofacture, as Marx put it—was really located in the eighteenth century.

J.D. Bernal, in *Science in History*, talks about a new scientific and technological revolution happening *simultaneously* from the beginning of the twentieth century. The twinning of science and technology in production was institutionalised through a variety of instruments—from publicly funded science and technical institutions to R&D laboratories that were a part of industry. Science began to be increasingly perceived as providing a competitive advantage to nations. France and Germany had set up their educational systems with this conscious purpose.

Science was perceived to be of direct importance to industrial production and had not only to be *produced* but also *reproduced*.

This system of reproducing science—continuously creating new scientists (or scientific workers)—engendered the twentieth century's complex of universities, institutions and research laboratories.

Bernal's key contribution in his seminal book, *The Social Function of Science*, was to recognise that the production of science and its reproduction required planning—science had now to be planned and funded.[29] He quantified the amount of scientific research (R&D) being done in the UK and was the first to measure the research intensity of each industry.[30] Finally, taking the example of the Soviet Union, he proposed an at least ten-fold increase in R&D expenditure for science in order to meet the demands of industrial production and larger social needs. In *The Social Function of Science*, Bernal also brought out how science, instead of being used for public good, was being misused for war and private appropriation. The enormous potential of science and technology was harnessed not to benefit the mass of people, but to increase profits for capitalists and supply the needs of war. Bernal felt that capitalism, with its intrinsic anarchy, was incapable of planning and utilising science to increase production—capitalism was a fetter on the growth of science as a productive force. He also felt that the Soviet system of planning science would give it a long-term competitive edge, not only in fulfilling the aspirations of the people but also in out-producing the capitalist mode of production.

In hindsight, it is easy to see that Bernal, like most other Marxists of his time, underestimated the resilience of the capitalist system. With the success of Big Science—typified by the Manhattan

[29] J.D. Bernal, *The Social Function of Science*, George Routledge and Sons, London, 1939. J.D. Bernal, *Science in History*, Vol. 1.

[30] 'Bernal, for example, was one of the first academics to conduct a national measurement of research in a Western country, although he used available statistics and did not conduct his own survey.' Benoît Godin, *The Making of Science, Technology and Innovation Policy: Conceptual Frameworks as Narratives, 1945-2005*, Centre Urbanisation Culture Société, Institut National de la Recherche Scientifique, p. 43, 2009.
https://espace.inrs.ca/id/eprint/9915/1/Godin_2009_412.pdf

Project—capitalism re-organised science very much on the lines that Bernal was suggesting. What went missing was of course the objective of public good. On the other side, Soviet science became bureaucratised and resistant to change—something Bernal had warned was a danger in the plan-based model of science.

Bernal's ideas on planning science were opposed by an influential group of scientists, including Michael Polanyi, the influential British-Hungarian polymath. He, along with others, argued that the planning of science was not possible as no one can predict the direction science may take.[31] They wanted freedom in science as opposed to planning—which they felt was very much a part of 'regimented' socialist thought. The post-war period saw the victory of Bernal's proposals for planned science, even in capitalist societies. And of his argument that countries needed to expand scientific research by a factor of ten as a percentage of GDP. Within two decades, this had become the norm in most advanced economies.[32] The irony is how quickly, in spite of the Cold War, capitalism adapted the planning model in science to the needs of war, while rejecting planning in all other areas of production. Bernal had shown that the Soviet Union, starting out as a backward economy, was able to develop by planning science and technology, building scientific/technology institutions and committing resources to both heads.

The example of the Soviet Union and Bernal's writings had a significant impact on developing countries, particularly India. Most developing countries saw the state as an instrument of development, and building scientific and technological infrastructure was seen

[31] Michael Polanyi, 'Rights and Duties of Science', in *Contempt of Freedom: The Russian Experiment and After*. London, Watts & Co., 1940 quoted from Robert E. Filner, 'Science and Marxism in England, 1930–1945', *Science and Nature*, No. 3, pp. 60–69, 1980. http://www.autodidactproject.org/other/sn-filner.html

[32] Christopher Freeman, 'The Social Functions of Science' in Francis Aprahamian and Brenda Swann (eds.), *J.D. Bernal: A Life in Science and Politics*, Verso, 1999.

as one of the key tasks of the state. In India, Bernal's work had enormous influence in the setting up of the Council of Scientific and Industrial Research and other scientific institutions—under Nehruvian policies towards science and technology. In fact, J.B.S. Haldane and Bernal both played important roles in the setting up of India's scientific and technology infrastructure.[33]

Bernal's *Social Function of Science* is also interesting for the way he looks at technology. He talks about how technology has to have a certain scale to make a difference to industrial production. Scientific and technological knowledge, as we have already seen, is universal labour. If labour produces capital, it also produces knowledge as a force for production. This knowledge may be considered, like capital, a form of 'dead' labour. Bernal's concept of what constitutes meaningful technology is very similar to Marx's understanding of capital. Marx says that money has to have a certain bulk before it can become capital. Similarly, Bernal held that technology must have a certain scale before it is meaningful. Obviously, both money and technology share this characteristic of capital.

However, knowledge of science and technology has one special property that distinguishes it from other forms of capital: money as capital or tangible capital (plant and machinery) depreciates with use. Knowledge as capital does not wear out but can be used again and again without loss. The issue of social control is relevant here, so that technology may be directed towards progress in the shape of public benefit and away from doing harm.

The destructive potential of science and technology has been apparent since the industrial revolution; modern warfare adds

---

[33] Veena Rao, 'J.B.S. Haldane: An Indian Scientist of British Origin', *Current Science*, Vol. 109, 3, 2015.
https://www.jstor.org/stable/24906123
See also Sabyasachi Chatterjee, 'Scientist as Revolutionary', *Frontline*, 25 May 2001.
https://frontline.thehindu.com/other/scientist-as-revolutionary/article30250584.ece

to the threat. Earlier, humanity could endanger its immediate environment but could not destroy civilisation along with much of the existing biosphere. Today, the destructive potential of runaway technology—nuclear weapons, climate change from greenhouse gases, biological weapons—can destroy the world as we know it. We need to recognise that the retreat from science and technology, towards anti-modernity, gained new impetus from the First World War. The use of poison gas—a product of modern science—preceded (and foreshadowed) the impact of atomic weapons in Hiroshima and Nagasaki in the Second World War.

This has added a new dimension to the issue of social control over technology and the democratisation of decision-making in major scientific and technological transitions. It is difficult to accept that only scientists and technologists, who understand their subject well, should be left fully in charge of all decisions concerning science and technology. We need to demystify policy decisions, often shrouded with the use of 'science' and 'expertise'. We need a people's science[34] along with a scientific/tech workers' movement to address this social task.[35]

INTELLECTUAL PROPERTY, KNOWLEDGE MONOPOLY AND THE RENT ECONOMY

The twentieth century saw the emergence of public-funded universities and technical institutions, while technology development was concentrated in the R&D laboratories of large

[34] T.V. Venkateswaran, 'Science for Social Revolution: People's Science Movements and Democratising Science in India', *JCOM Journal of Science Communication,* Vol. 19, 6, 2020; https://doi.org/10.22323/2.19060308
Science for the People.
https://scienceforthepeople.org/
[35] The free software movement was conceived to play this role. But with major tech companies adopting the free software model, the movement needs to also look at the challenge of new tech monopolies and how tech workers can fight such monopolies.

corporations. The age of the lone inventor—Edison, Siemens, Westinghouse, Graham Bell—had ended with the nineteenth century.[36] The twentieth century was more about industry-based R&D laboratories, where corporations gathered together leading scientists and technologists to create the technologies of the future. In this phase, capital was still expanding production. Even though finance capital was already dominant over productive capital, the major capitalist countries still had a strong manufacturing base. In this phase of development, science was regarded as a public good and its development was largely concentrated in the university system or publicly funded research institutions. Technology development was largely regarded as a private enterprise. Science was supposed to produce new knowledge, which could then be mined by technology to produce artefacts.[37] The role of innovation was to convert ideas into artefacts. The system of intellectual property—patents and other rights—arose to provide protection to the useful ideas embodied in artefacts. From the beginning, patents also had a public purpose—the state-granted monopoly for a certain period was meant to ensure the eventual public disclosure of the invention: the quid pro quo being full public disclosure in lieu of a limited-term monopoly.

The transformation of this system that had existed for several centuries came about as a result of two major changes in the production of knowledge. The first relates to the way in which, under the neoliberal order, the university system of knowledge-production has been transformed into a profit-making commercial enterprise.[38] Secondly, the distinction between science and

[36] Philo T. Farnsworth, the inventor of television, has been described as the last lone inventor. He fought against the corporate power of David Sarnoff, and RCA, the most powerful company in the business of broadcasting. It was a David versus Goliath battle. Evan I. Schwartz, *The Last Lone Inventor: A Tale of Genius, Deceit, and the Birth of Television,* Harper Collins, 2002.

[37] I have dealt with the relationship between science and technology in more detail in the essay 'Restoring Conceptual Independence to Technology'. See Chapter 5 in Section II of this book.

[38] 'Academic administrators increasingly refer to students as consumers and to

technology has blurred considerably and the two are more closely integrated than before. For example, an advance in genetics can almost seamlessly lead to an artefact—a drug, a diagnostic tool or a seed—that is both patentable and marketable. Similar is the case of innovations in the field of electronics and communications. Many disciplines of science and also research output in universities, are, in consequence, driven closer to the systems of production. The conversion of the university system into a system producing knowledge directly for commercial purposes has happened in tandem with the destruction of the R&D laboratories that were so much a part of the industrial landscape of the twentieth century. Finance capital controls university science, not just through 'investment' in R&D, but also the purchase of 'knowledge'. Its monopoly is exercised through buying the patents that university research produces. This monopoly in turn allows finance capital to dominate over industrial capital.

The end of the twentieth century revealed the rupture of finance capital and productive capital. Today, global capital operates far more as disembodied finance capital, controlling production at one end with its control over technology and markets at the other. In this phase, where capital increasingly lives off speculation and

---

education and research as products. They talk about branding and marketing and now spend more on lobbying in Washington than defence contractors do'. Jennifer Washburn, *University, Inc.: The Corporate Corruption of Higher Education*, Basic Books, 2005. Before the Bayh-Dole Act 1980, any invention or knowledge produced with public money had to remain in the public domain to be used by any individual or company. The universities have a vested interest in patenting—for example medicine—selling the rights to the private sector and generating either large one-time payments, or a stream of royalties or both. 'This report shows that NIH funding contributed to published research associated with every one of the 210 new drugs approved by the Food and Drug Administration from 2010–2016'. Ekaterina Galkina Cleary, Jennifer M. Beierlein, Navleen Surjit Khanuja, Laura M. McNamee, Fred D. Ledley, 'Contribution of NIH Funding to New Drug Approvals 2010–2016' *Proceedings of the National Academy of Sciences*, Vol. 115, 10, pp. 2329–2334, 2018.
https://doi.org/10.1073/pnas.1715368115

rent, there is also a marked separation of knowledge as capital from productive or physical capital—plant and machinery. Foxconn/Hon Hai Precision Industries manufactures Apple products but cannot claim a major share in the profits from their sale, since Apple holds the intellectual knowledge and property rights. Roughly, Apple gets 31 per cent of the profits from an iPhone sale, Foxconn less than two per cent.

The transformation of capital to rent seeking, by using its monopoly over knowledge—patents, copyrights, industrial designs, etc.—characterises the current phase of capital. With this, the advanced capitalist countries have increasingly become rentier and 'service' economies. In essence, they dominate the world by virtue of controlling the global financial structure, new knowledge required for production, and distribution through retail and global brands.

Even as universities are captured by capital and turned into what is termed as University Inc, the new knowledge they produce is still publicly funded.[39] This is true alike of advanced capitalist countries and those like India. The direction of scientific research is dictated by private capital, which takes over any successful outcome, and yet this transformation of science did not come about through being privately funded. The cost of fundamental research is high and only a few of its research outputs may have immediate benefits in terms of advancing technology. This is where the state, whether in electronics or in genetics, takes care of the costs while the patents are handed over to private capital. A hallmark of the neoliberal system is the socialisation of risk and privatisation of rewards.

The understanding that science needs to be restored as an open and collaborative exercise has given birth to the commons movement. By a curious sleight of hand, capitalism sees the finite commons—the atmosphere and large water bodies such as lakes, rivers and oceans—as infinite, and demands the right to dump

---

[39] Washburn, *University, Inc.*.

waste in these commons. Yet it regards knowledge, capable of being copied infinite number of times without loss, as finite and demands monopoly rights over it!

Never before has society had the ability it does today to bring together different communities and resources in order to produce new knowledge. It is social, universal labour, and its private appropriation as intellectual property under capitalism stands in the way of liberating the enormous power of the collective to generate new knowledge and benefit people.

SOCIAL CONSTRUCTIVISM AND TECHNOLOGICAL DETERMINISM

Certain debates have gained ground in what is called science, technology and society studies, or social studies of science, or science policy studies. They deal with how society shapes the way technology develops and how technology in turn shapes society. One particular position that has gained ground is social constructivism.

Social constructivism is an influential school of scholars arguing that technology does not unfold in a unilinear fashion by some inner logic, but is the result of a series of social choices. If this was all they asserted, we would have little to quarrel about. The key problem in their account is that the social, used merely as a sociological term, is devoid of class and industrial production. Any success in advancing production is treated as one factor on par with other factors—such as a preference among young men for speedier bicycles. In this view, technological effectiveness in solving a given problem is only one driver among many others. The second problem with social constructivism is that it assumes that technology is infinitely plastic—it can offer a range of solutions of almost equal technological effectiveness.

Social constructivists use the principle of neutrality between different technology options. In this scheme, any cultural preference or economic need is treated as if each choice had equal weight. The role of class, or class interest, in promoting a certain kind of technology disappears in such a paradigm. Russell makes this point in his criticism of social constructivism, 'If we accept that arguments over technological options are socially constructed, then it follows that a relativist approach with respect to them leads us into relativism with respect to social interests—in other words, political neutrality'.[40] It says nothing about the impact of the choices being made and makes no distinction between an emancipatory technology and an oppressive one.[41] That there are social choices being made in the development of technology does not tell us anything about the choices that should be made from an emancipatory viewpoint: it provides no instrument of intellectual struggle. The second problem with the social constructivists' account stems from their belief that as technology is infinitely plastic, there are infinitely many choices possible in developing technology, with the final choice depending on largely sociological factors. On the contrary, the feasible designs for most technologies are strictly limited and social shaping drives artefacts only within a narrow range of design choices.[42] Social choice from within a few technology options that have already been sieved through as technologically 'effective' is quite different from what social constructivism perceives as the driver of technology, or in its language the 'construction' of technology.

In the larger context, social constructivism would belong to

[40] Stewart Russell, 'The Social Construction of Artefacts: A Response to Pinch and Bijker', *Social Studies of Science,* Vol. 16, 2, pp. 331–346, 1986. https://www.jstor.org/stable/285210

[41] Langdon Winner, 'Upon Opening the Black Box and Finding it Empty: Social Constructivism and the Philosophy of Technology', *Science, Technology, & Human Values,* Vol. 18, 3, 1993. https://www.jstor.org/stable/689726

[42] Walter G. Vincenti, 'The Technical Shaping of Technology: Real-World Constraints and Technical Logic in Edison's Electrical Lighting System', *Social Studies of Science,* Vol. 25, 3, 1995. https://www.jstor.org/stable/285506

Science in the Light of Social History

the externalist view of the development of science and technology. Marx, Engels, and Marxists such as Bernal have been accused of both: a completely externalist view—where only external social factors drive science and technology—as well as technological determinism. In Bernal's case, his critics argue that *Science in History* is a crude externalist account of the history of science, meaning that social factors create needs, and science is developed to meet these needs. This was also the criticism of Boris Hessen's paper on Newton at the 1931 Congress that we referred to earlier. Bernal's *Social Function of Science* is viewed to be its opposite, an exercise of technological determinism.

What does technological determinism mean? It means that technology creates society—it is relatively autonomous of society and creates a society consistent with the prevailing level of technology.

Let us take the first part of technological determinism—that technology development creates social change. There is no doubt an underlying belief in Marxism that as technology and science develop, this leads to the development of productive forces and their conflict with existing relations of production. However, all serious Marxists have also shown how the system of production of science and technology is itself a product of the existing production relations. Though a closed loop, it is not a static one. Even if the social monopoly of certain classes tries to keep technology static, it will change over time, albeit slowly, and this will introduce changes in the system of production. But technology is as much determined by society as society is by technology. The relationship of technology and society is a dialectical one: a two-way, dynamic relationship.

The argument that the kind of technology that arises is imprinted with the class relations of society is different from the same argument vis-à-vis science. The products of technology have social functions and therefore do encode societal relationships in some way. Marx says, 'The hand-mill gives you society with the feudal lord; the steam-mill society with the industrial capitalist'.[43] Here, 'gives' should be understood as 'being coterminous with', not as cause and effect. Let us take one example. A labour shortage economy may have a high cost of labour and need to automate for improving quality and reducing labour costs. A labour surplus economy may have low labour costs and find such automation not competitive. Here, a selective automation that ensures quality in the production process without automating the rest—a form of semi-automation—may be the preferred technology choice. Therefore, technologies are far more located within the social context.

Let us take another example of how social relations shape technology. One of the reasons that hybrid seeds were favoured during the green revolution was because hybrid seeds do not breed true, therefore the next generation of crops from the seeds of the present one would lose the very properties that had led the farmer to purchase hybrid seeds. If the seeds bred true, the farmers would control their production after a one-time purchase. The hybrids were consciously created to become a monopoly of the seed companies, so that farmers would have to buy the same seeds again and again. The decision to create hybrid seeds that do not breed true was the result of this class choice.[44] Bt cotton in India is promoted through hybrids so that Monsanto can retain control over the seed market.In this larger sense, technology does reflect social relationships, just as society reflects the prevailing level of technology. Given a technology, we can see the kind of society that created it. Technological determinism makes it out to be a one-

---

[43] Karl Marx, 'The Metaphysics of Political Economy' in *The Poverty of Philosophy*, 1847.
http://www.marxists.org/archive/marx/works/1847/poverty-philosophy/ch02.htm#s2

[44] Richard C. Lewontin and Jean-Pierre Berlan, 'The Political Economy of Hybrid Corn', *Monthly Review*, 1 July 1986.
https://monthlyreview.org/1986/07/01/the-political-economy-of-hybrid-corn/

way process, while Marxists look upon the process of technology creating society and society creating technology as a dialectical, two-way process.

THE FUTURE DIRECTION OF SCIENCE

When Marx spoke of science allowing humanity to 'control nature', he was drawing a contrast with how nature had earlier dominated humanity, and how only science and technology liberated humanity from blind obedience to nature. This is different from the view that science and technology—most such views collapse them as one entity—have an ideological underpinning of domination over nature. In this view, knowing nature and domination over nature are identical and this amounts to the ideology of science: an ideology of domination now endangering the planet. If this view is to be followed through, retreating from science is the only way to save the world. Much of the anti-science and anti-technology views share this viewpoint, if couched in different terms. The 'enemy' is science, technology, development, and not capital. It is an outlook divorced from the class issue of who owns the means of production and uses them to what end.

The problem with retreating from science and (implicitly) from development is that the world does not stand still. At any point of time society is in a state of dynamic equilibrium with nature. It cannot opt out of this equilibrium and seek stasis—it either goes forward or inevitably backward. Keeping the current capitalist system in stasis implies that a minority will be rich, while major sections of the world's population will be at subsistence levels. Or, if not this, we and the planet soon run out of resources for survival. The argument that we should retreat from science is not only a retreat from the hope of building a better society, it is also the refusal to understand that without developing knowledge and our tools of production, we cannot even stay at our current levels of development. Or provide society with the redistributive justice required in the world.

Let us take two examples. One is medicine. It is a simple evolutionary fact that bacteria become immune to specific antibiotics over time. What happens if we stop developing new medicine? Our armoury of medicines will become ineffective over time. We will still be able to use them by cycling different antibiotics, but their effectiveness will be nowhere near today's level. Apart from this, we have new diseases that arise—HIV/AIDS is one example. Or SARS-CoV-2/COVID-19. New medicines are not a mere fancy of global corporations and their scientists but a vital requirement.

It is always possible for our defence mechanisms to evolve and make us relatively immune to a disease. It has happened before. The problem is, this may take a long time and can devastate societies. Plus, human volition alone cannot generate immunity. Plague and Black Death are not distant enough in our past to make it seem a simple process! Nor did such a course eliminate polio or small pox. It was only with vaccines that these diseases could be contained or eradicated.

The other example is food production. It is estimated that the human population was about 300 million 2,000 years back, and took about 1,600 years to double.[45] The next doubling took only 250 years—by 1850, the population had reached 1.2 billion. By 1950, another 100 years, we had crossed 2.5 billion, and reached more than 6 billion in 50 more years. Malthus had predicted that humanity would run out of food since our population increases geometrically and food production increases linearly. Obviously, scientific and technological change has kept pace with the population and, therefore, so has food production, defeating the

---

[45] United Nations Department of Economic and Social Affairs, Population Division (2021). Global Population Growth and Sustainable Development. UN DESA/POP/2021/TR/NO. 2.
https://www.un.org/development/desa/pd/sites/www.un.org.development.desa.pd/files/undesa_pd_2022_global_population_growth.pdf

Malthusian paradigm. But what happens if we freeze science and technology? Simply that food production would no longer keep pace with an increased population, or even the stabilised population predicted in the next 50 years.

The issue of what science can do and what science actually does is a terrain of struggle within science and also in the larger social arena. It is a battle for the future direction of science, and part of a larger battle for the allocation of resources for the betterment of humanity. It is a battle against irrationalism in science and in society. It needs that socially conscious scientists, working people, and scientifically conscious progressive sections of society should wage this battle together. It also means the democratisation of scientific decision-making; a few scientists cannot sit and decide what society needs.

The struggle waged by Bernal and others for social responsibility among scientists, came from this perspective, as did the many associations of scientists for peace and against nuclear weapons.[46] This is also why the popularisation of science was and continues to be a major thrust of progressive science movements. It is impossible to freeze the boundaries of knowledge, even in select areas. We may artificially create boundaries in terms of disciplines, but in nature all knowledge is inter-connected. One cannot unravel nature at one end without unravelling it elsewhere.

We are not arguing for a simple use/misuse model of science—that science is value neutral and it is either used or misused by social groups/classes/nations. The scientist or the scientific worker is not just creating neutral knowledge—s/he is developing knowledge which may be used for human or environmental destruction, or to favour the few. One cannot divorce the two—creating scientific

---

[46] The Association of Scientific Workers and the World Federation of Scientific Workers were movements that sprang up after the Second World War. The people's science movement in India and science for the people in the US were products of the 1960s–1970s turmoil centred on the anti-war movements that also affected the scientific community. Another obvious example is the Pugwash Conference that was founded to fight against nuclear weapons.

knowledge and fighting at the level of society. Both of these must go together in a struggle to reshape science to serve the needs of larger humanity. The choice of problems, the kind of solutions offered, the direction that science takes are all a part of this larger struggle.

Helena Sheehan encapsulates this position on science, technology and society:

> In the tradition of Bernal, the left took its stand with science. I do not believe that the debunking of science in terms of its cognitive capacity or its social potential is an appropriate activity for the left. It is neither epistemologically sound nor politically progressive. The left should take its stand with science, a critically reconstructed, socially responsible science, but with all the higher possibilities of science. It should engage in a radical critique of the incorporation of science to global capital. It should open a path to the progressive potentialities of science.[47]

This is the challenge we face today.

---

[47] Helena M. Sheehan, 'J.D. Bernal: Philosophy, Politics and the Science of Science', *Journal of Physics: Conference Series*, Vol. 57, pp. 29–39, 2007. https://iopscience.iop.org/article/10.1088/1742-6596/57/1/003

8. The Untold Story of the Left in Indian Science

The story of modern Indian science generally focusses on Nehru and the scientific institutions that were built to help India industrialise. This story is incomplete, as it does not take into account the contributions of scientists from the Indian left, such as Meghnad Saha, Sahib Singh Sokhey, and Syed Husain Zaheer, as well as figures from the international left such as J.D. Bernal and J.B.S. Haldane, who are very much at the core of the story.

Science in India was anchored in the vision that a modern nation could only be built by combining a scientific outlook—what Nehru called 'scientific temper'—with a significant role for state planning in science, technology, and the economy. That science can be planned may be a commonplace view today, but was completely heretical in the scientific community before the Second World War. The 1917 revolution with its use of planned industrial development, harnessing science to build a modern nation, was anathema to the established order in the scientific community. The Soviet delegation to the 1931 International Congress of the History of Science in London, led by Bukharin, provided a glimpse of what a different vision of science in society could be. A set of brilliant, young British scientists—J.D. Bernal, J.B.S. Haldane, Joseph Needham, Lancelot Hogben—were inspired by the Soviet papers at the meet. This led to Bernal's famous work, *The Social Function of Science* (1939), the formation of the science and society movement, and the unionising of scientists as scientific workers.

After the Second World War, the science movement gave rise to the World Federation of Scientific Workers, and also became an important component of the global peace movement—the World Peace Council. Bernal and Haldane had close connections to Indian science subsequently, while Needham worked with his Chinese colleagues on the monumental *Science and Civilisation in China*: fifteen books in the series were published in his lifetime, and the project has continued since, running to twenty-seven books in seven volumes today. All three scientists, Bernal, Needham, and Haldane, were either members of the Communist Party of Great Britain or very close to it; Needham's biographer Maurice Goldsmith claims that Needham, while close to the CPGB, was never a member.

In Calcutta during the 1930s, a group of young scientists with Meghnad Saha were similarly engaged: studying science, planning, and the socialist experiment in the Soviet Union. It was Meghnad Saha who influenced Subhas Chandra Bose, when he became the president of the Indian National Congress, to set up a planning committee. The Planning Committee was set up by Bose under Nehru's leadership and created the outlines of the eventual plan for India's industrialisation and its new institutions of science and technology. After independence it became the Planning Commission, and provided the basis for expanding India's science and technology infrastructure. The expansion of the Council of Industrial and Scientific Research with its laboratories, atomic energy, space, and educational institutions, including the Indian Institutes of Technology, was an outcome of this vision of planning and marked the importance of science for a self-reliant India.

Nehru was aware that if India was to industrialise, the state had not only to play an active role in the economy but also build the scientific institutions that could support such a path. In his quest to build scientific institutions, he invited Bernal to India a number of times. The geneticist J.B.S. Haldane settled down in India, even taking Indian citizenship. Haldane could not negotiate India's bureaucratic science establishment: he told Nehru that the Council of Scientific and Industrial Research (CSIR) ought to be renamed as the Centre for Suppression of Independent Research.[1]

---

[1] S. Irfan Habib, 'Legacy of the Freedom Struggle: Nehru's Scientific and Cultural Vision', *Social Scientist*, Vol. 44, 3/4, pp. 29–40, 2016.

A number of scientists in India, such as Meghnad Saha, Husain Zaheer and Sahib Singh Sokhey, were not only the founders of post-independence Indian science, but also close to the Communist Party of India.[2] They were included in the roll of honour of US scholars, as 'fellow travellers', a term that became popular in the infamous McCarthy era. Meghnad Saha, along with Homi Bhabha, laid the foundation for nuclear physics in the country, and Husain Zaheer, as director general of the CSIR, expanded it significantly.

Sokhey, appointed as the first Indian director of the Haffkine Institute in 1932, built it into a premier research institution that could also produce vaccines and medicines at an industrial scale. Even though he was a colonel in the British Indian Army, he headed the health section of the Planning Committee set up in 1938 by the Indian National Congress, with Jawaharlal Nehru as its chairman.[3] After independence, despite facing opposition, Sokhey laid the foundation of two public sector units that began India's journey towards self-reliance—Indian Drugs & Pharmaceuticals Ltd and Hindustan Antibiotics Ltd.[4] He also became president of

---

https://www.jstor.org/stable/24890242

[2] Sokhey headed the All India Peace and Solidarity Organisation affiliated to the World Peace Council.

[3] Sokhey Committee Report, 1948.
https://www.scribd.com/document/25959490/Sokhey-Committee-Report-1948

[4] Despite Nehru's backing, the Western MNC lobby, aligned with sections of the Indian government, bitterly opposed the setting up of public sector drug companies. Sokhey had to bring in the blueprint of an antibiotic plant that he and his associate, K. Ganapathi, had already designed, with support from the UNICEF/WHO. Again, it was Sokhey who was instrumental in bringing Soviet technology to IDPL. For more details on this bitter battle, see Nasir Tyabji, 'Gaining Technical Know-How in an Unequal World: Penicillin Manufacture in Nehru's India', *Technology and Culture*, Vol. 45, 2, pp. 331–349, 2004.
https://www.jstor.org/stable/40060744
Harkishan Singh writes that Sokhey understood that self-reliance in medicines meant building a self-reliant chemical industry. Harkishan Singh, 'Sahib Singh Sokhey (1887–1971): An Eminent Medico-Pharmaceutical Professional', *Indian Journal of History of Science*, Vol. 51.2.1 pp. 238–247, 2016.

the All India Peace Council (a part of the World Peace Council), and later on, president of the Association of Scientific Workers of India (ASWI), the Indian affiliate of the World Federation of Scientific Workers.[5] Jawaharlal Nehru was the first president of the ASWI in 1947. The World Federation of Scientific Workers (WFScW) was founded in July 1946 in London, at the initiative of the ASW—Frédéric Joliot-Curie, a Nobel Prize-winning nuclear physicist, was its first president.[6] Joliot-Curie was dismissed in April 1950 from his position as French High Commissioner for Atomic Energy for being a communist.[7] The WFScW defined itself as a science and society movement and not as a trade union.

Interestingly, Homi Bhabha was a participant in the founding conference.[8] Joseph Needham, who headed UNESCO's science division, chaired the International Scientific Commission for the Investigation of the Facts Concerning Bacterial Warfare in Korea and China. The commission indicted the US for war crimes in its

---

https://doi.org/10.16943/IJHS%2F2016%2FV51I2%2F48435

[5] The Association of Scientific Workers of India (ASWI) was founded in 1918 on the lines of the Association of Scientific Workers (ASW) in the UK (originally called National Union of Scientific Workers in the United Kingdom). It was at first a trade union but later included major scientific figures, not just those who identified themselves as Marxists, but also scientists such as Lancelot Hogben who were in the Labour Party.

[6] J.G. Crowther, 'The World Federation of Scientific Workers', *Nature*, 160, pp. 628–629, 1947.
https://www.nature.com/articles/160628a0

[7] Michael Clark, 'French Remove Joliot-Curie as Chief for Atomic Energy; Communist Head of Science Commission Dismissed by Premier Bidault FRANCE DISMISSES RED ATOMIC SAVANT Scientist Back in Red Favor Colleagues Support Savant', *New York Times*, 29 April 1950.
https://www.nytimes.com/1950/04/29/archives/french-remove-joliotcurie-as-chief-for-atomic-energy-communist-head.html

[8] Petitjean writes, 'Among the Participants were Leon Rosenfield from Netherlands, Homi Bhabha from India and Eric Burhop from Australia.' Patrick Petitjean, 'The Joint Establishment of the World Federation of Scientific Workers and of UNESCO after World War II.' *Minerva*, 46, 2, pp. 247–270, 2008.
http://www.jstor.org/stable/41821462

1952 Report.⁹ Sokhey had agreed to join the commission, but the Government of India asked him not to. For a long time, the US denied that it had carried out attacks with bioweapons against the Koreans and the Chinese. Declassified documents from the US archives now confirm many of the events that the Needham Commission had described.¹⁰ The Japanese Unit 731 had practised biological warfare experiments on Chinese and Allied prisoners of war, killing 3,000.¹¹ The unit's leaders were provided full protection against war crimes in lieu of turning over all their 'research' to the US.¹² It is their 'research' that kick-started the infamous bioweapons research centre, Fort Detrick, and was the source of the chemical and biological weapons used by the US in the Korean war.

---

⁹ Jeffrey Kaye, 'The Long-Suppressed Korean War Report on U.S. Use of Biological Weapons Released At Last' *Insurgent Intelligence*, 20 February 2018. https://mronline.org/2018/02/21/the-long-suppressed-korean-war-report-on-u-s-use-of-biological-weapons-released-at-last/#edn7

¹⁰ See two articles by Jeff Kaye: 'CIA Document Suggests US Lied about Biological, Chemical Weapon Use in the Korean War', *Shadow Proof*, 10 December 2013. https://shadowproof.com/2013/12/10/cia-document-suggests-u-s-lied-about-biological-chemical-weapon-use-in-the-korean-war/ and '"A Real Flood of Bacteria and Germs": Communications Intelligence and Charges of U.S. Germ Warfare during the Korean War', *Counterpunch*, 3 September 2021. https://www.counterpunch.org/2021/09/03/a-real-flood-of-bacteria-and-germs-communications-intelligence-and-charges-of-u-s-germ-warfare-during-the-korean-war/

¹¹ The figure 3,000 is an underestimate that emerged from the cross examination of Japanese scientists. Another estimated 4,000 civilians were killed by infectious diseases such as cholera and anthrax when packages were dropped in bombs on Chinese citizens in villages. Thomas Powell, 'Biological Warfare in the Korean War: Allegations and Cover-Up', *Socialism and Democracy*, Vol. 31, 1, pp. 23-42, 2017. https://doi.org/10.1080/08854300.2016.1265859

¹² Eamonn Fingleton, 'Imperial Japan's Abominable Dr. Death, and the Most Disgraceful War Crime "Amnesia" In History', *Forbes*, 9 March 2014. https://www.forbes.com/sites/eamonnfingleton/2014/03/09/imperial-japans-abominable-dr-death-and-the-most-disgraceful-war-crime-cover-up-in-history/?sh=48828a3475de

SCIENCE AND THE LEFT IN INDEPENDENT INDIA

The left's contribution to India's science infrastructure after independence was not restricted to expanding science and research institutions but included the fight for self-reliance. This could not happen without a bitter tussle with the deeply entrenched interests of multinationals and the colonised mindset of a number of scientists and political leaders, a battle that continues today. The Indian left played a role in building national science institutions, integrating them with the larger battle for self-reliance and building a scientific outlook. The left also contributed to advances in science.

I would like to focus on one area, the pharmaceutical industry, to bring out certain aspects of the larger story of the left's contributions to Indian science and technology.[13] When the British left, the pharmaceutical industry was completely in the hands of British owners, who produced the active pharmaceutical ingredient (API) in the UK and only packaged it here for sale. While there were small Indian pharma companies, they lacked the backing of scientific research or a plan for how to fight the British-era Indian patent law and the legal monopoly it had instituted, which multinationals now enjoyed.

It was a two-pronged battle: to change the patent laws in the interest of the Indian people, while also building the scientific infrastructure and know-how required by the indigenous drug industry.

As the first Indian director of the Haffkine Institute, Sahib Singh Sokhey converted it from what was a cottage industry for producing vaccines into a fully modern facility. This core team later empowered India's fledgling public sector units set up with Soviet and WHO help—Hindustan Antibiotics Ltd. and Indian Drugs and Pharmaceuticals Ltd. Sokhey was clear that the

---

[13] Prabir Purkayastha et al (eds.) *Political Journeys in Health, Essays by and for Amit Sengupta*, LeftWord Books, 2020.

The Untold Story of the Left in Indian Science

Indian Patents Act needed to change if India was to develop its indigenous pharmaceutical industry. (Unfortunately, this part of the Sokhey Committee Report had to wait more than two decades to be enacted as the Patents Act 1970.)[14] Sokhey knew that the challenge before the emerging pharmaceutical industry was threefold: a) India needed scientific knowledge to produce existing drugs as well as a new generation of drugs; b) India needed the ability to produce such drugs at an industrial scale, not just in the laboratory; and c) the drugs produced had to be cheap enough to make them accessible to the Indian people. People today may have forgotten that the life expectancy of Indians at the time of independence was 32 years, cut short by infectious diseases, epidemics, and malnutrition.

The Patents Act was one of two major issues for pioneers such as Sokhey. The other issue was setting up public sector units to produce drugs, and developing research institutions that would help make India self-reliant.

The laboratories of the Council of Scientific and Industrial Research (CSIR) created the scientific and technological knowledge infrastructure required for an indigenous Indian pharmaceutical industry. It was the CSIR infrastructure, created under the leadership of Husain Zaheer, Nitya Anand—the director of the Central Drug Research Institute (CDRI), Lucknow—and the National Chemical Laboratories, Pune, that made it possible for the Indian drug industry to break the stranglehold of Western MNCs on the Indian market.

There are several figures of interest—besides Sokhey—during this transition period of the pharmaceutical industry. One is Khwaja Abdul Hamied, the founder of CIPLA, a follower of Gandhi and an ardent nationalist.[15] Along with Sokhey, Hamied was a member

[14] National Planning Committee Series, Report of the Sub-Committee on National Health, Chairman, Col. S. S. Sokhey, 1947. https://ruralindiaonline.org/en/library/resource/national-planning-committee-series-report-of-the-sub-committee-national-health/

[15] CIPLA was founded in Mumbai by Khwaja Abdul Hamied as The Chemical,

of the committee for the expansion of the CSIR and continued to be associated with the Council.[16] His son, Yusuf Hamied, followed his father's philosophy of cheap medicines for the people: Cipla, sometimes described as the Robin Hood of drugs, would go on to provide affordable AIDS drugs to people in developing countries.[17] Another figure of note is Dr Nitya Anand; as the director of CDRI, he pioneered research into the new processes required to make pharmaceutical products for the market. He later chaired (along with S.P. Shukla) the National Working Group on Patents Law which B.K. Keayla had helped set up in 1988.

We, activists with the Delhi Science Forum since 1978, knew the story of 'later pioneers' such as Nitya Anand, Abdul Hamied, or Ranbaxy's founder Bhai Mohan Singh.[18] However, we did not know the legacy of the earlier generation of activists—progressives and anti-imperialists, a part of the independence movement—whose struggle we had inherited.

Soon after independence, a committee chaired by Justice Bakshi Tek Chand, formerly of the Lahore High Court, was set up to examine changes to the Patents Law. The Committee observed that the colonial-era law had led to high cost of medicines, but it failed to come up with an alternate framework. India had used the public sector route to develop the indigenous manufacture of antibiotics in the 1950s, but the bulk of medicines in the Indian market was still in the hands of multinationals.

In 1957, a committee chaired by a retired judge of the Supreme

---

Industrial & Pharmaceutical Laboratories, in 1935. The name of the company was changed to Cipla Ltd. on 20 July 1984.

[16] Harkishan Singh, 'Sahib Singh Sokhey'.
[17] Goldapple, 'India's Robin Hood of Drugs', *Breakthrough Briefs*, 19 September 2016.
http://breakthrough.unglobalcompact.org/briefs/cipla-indias-robin-hood-of-drugs-yusuf-hamied/
[18] After Bhai Mohan Singh handed over Ranbaxy to his son, Parminder Singh, Ranbaxy switched sides. Parminder Singh decided that partnership with multinationals on patented drugs and manufacturing generics made for a better business strategy for Ranbaxy.

Court, Justice Rajagopala Ayyangar, was set up to suggest the way forward on patents. In his report, Ayyangar acknowledged the role of Sokhey's close associate, K. Ganapathi, in understanding the implications of patents for the pharmaceutical industry. While Ayyangar did not go as far as Sokhey wanted—abolishing patents altogether—he did suggest changes to remove product patents in the food, drugs and chemical sectors.[19] Instead, he suggested granting only process patents in these sectors and restricting them to a shorter period. A draft bill based on the Ayyangar Committee Report was first introduced in Parliament in 1965. It was then referred to a joint parliamentary committee which submitted its report in 1966, and was finally passed by Parliament in 1970.[20] In the post-independence period, MNCs continued to keep their manufacturing capacity in the country at a minimal level, and used their control over new drugs[21] to maintain super profits,[22] the bulk of which flowed back to the parent companies.[23] The price of

[19] Shri Justice N. Rajagopala Ayyangar, 'Report on the Revision of the Patents Law', September 1959.
https://ipindia.gov.in/writereaddata/Portal/Images/pdf/1959-_Justice_N_R_Ayyangar_committee_report.pdf

[20] The Patents Bill, 1965, Report of the Joint Parliamentary Committee, November, 1966.
https://eparlib.nic.in/bitstream/123456789/755572/1/jcb_03_1966_patents_bill.pdf#search=null%20Joint/Select%20Committee%20on%20Bills%20[1960%20TO%201969]%201966
Though the Act was passed in 1970, it became operative with the Patents Rule being notified in April 1972.

[21] Gehl Sampath, 'Innovation in the Indian Pharmaceutical Industry' in *Economic Aspects of Access to Medicines After 2005: Product Patent Protection and Emerging Firm Strategies in The Indian Pharmaceutical Industry*, United Nations University-Institute for New Technologies, 2005.
https://www.researchgate.net/publication/42795517_Economic_Aspects_of_Access_to_Medicines_After_2005_Product_Patent_Protection_and_Emerging_Firm_Strategies_in_the_Indian_Pharmaceutical_Industry

[22] The Report of the Joint Parliamentary Committee on the Patents Bill (1966) gives various examples of MNCs exploiting their patent monopolies to charge exorbitant prices.

[23] The Hathi Committee Report, Sudip Chaudhuri, and P.G. Sampath have all written about the control of the MNCs over the Indian pharmaceutical market.

medicines was high and out of reach to most Indians. The British India Patents Act of 1911 should have been changed soon after independence. But this actually took more than two decades—an indication of the strength of the multinationals as well as the neo-colonial lobby in India.[24]

It was obvious that patents held by MNCs meant expensive medicines, and less access to medicines for most Indians. Then why—and how—did this delay in changing patents law take place?

BEHOLD THE FLAG-BEARER OF THE 'FREE WORLD'

In countries like India, the battle over pharmaceuticals was as much a battle for self-reliance as for affordable medicine. But the West viewed this attempt to build a self-reliant industry, free of multi-national control, as an alignment with communism. This was also how the West perceived non-alignment. Indeed, with control over knowledge becoming something of a strategic battle with the 'reds', the pharma companies considered themselves Cold War warriors. In his testimony to the Kefauver Committee's Congressional hearings on the drug industry, the president of Merck & Co, J.T. Connor, spoke of how Merck, a Big Pharma player, had allegedly won 'an initial skirmish with the Soviet Union in India last year, as Merck and the Soviet government fought for the right to establish a manufacturing plant in India'.[25] This is a reference to Sokhey's attempts to bring antibiotic technology to India with the help of the World Health Organisation, UNICEF, and the Soviet Union.[26] Connor went on to claim that 'our industry has grown into a significant national asset, these daily contributions to the war against disease are well known . . . . but [our] potential

---

[24] Tyabji, 'Gaining Technical Know-How in an Unequal World'.
[25] Dominique A. Tobbell, '"Who's Winning the Human Race?" Cold War as Pharmaceutical Political Strategy', *Journal of the History of Medicine and Allied Sciences*, Vol. 64, 4, pp. 429–473, 2009. https://doi.org/10.1093/jhmas/jrp012
[26] Tyabji, 'Gaining Technical Know-How in an Unequal World'.

contributions to the world struggle against communism are only beginning to become apparent'.

Merck's CEO positioned Big Pharma's battle to maintain a strong patents regime as a global battle against the Soviet Union. Big Pharma was fully aware that any country with a reasonable industrial base would be able to manufacture drugs and pharmaceutical products. Hence patents for new drugs were crucial for Big Pharma's global monopoly. And India was not just another market. The country was developing its science institutions; it had built up capacity in the cutting-edge technology of the day, antibiotics; and it had a huge internal market. The stakes went far beyond India: this is why the battle continued for more than two decades before India could change its patent laws. India's success was a case of fears-come-true for global capital—its global market was truly being endangered.

The Kefauver Committee's Report, submitted in 1961, pointed out that the US drug companies charged as much as 7,000 per cent of their costs. And who was charged this highest price in the world?[27] The poorest people, the people of India. This made big news in India, and gave a further fillip to the demand to change India's patents laws.

The Kefauver Report suggested limiting product patent monopoly to three years,[28] pointing out (as the Ayyangar Committee had also done), that advanced European countries such as Germany, Switzerland, Italy and France did not grant product patents, only process patents.[29] The Report also suggested that

[27] Judiciary Sub-Committee on Anti-Trust and Monopoly, 87th Cong. 1st Session., Rep. No. 448 (27 June 1961), showing India with the highest prices of the seventeen countries surveyed, which included the United States.
[28] Kefauver's Report would have died a natural death since it was bitterly opposed by the pharmaceutical industry and large sections of the US political establishment. But the thalidomide tragedy resuscitated it—as the Kefauver-Harris Amendments to the Federal Food, Drug, and Cosmetic Act that demands proof of the efficacy of a new drug before granting it a patent.
[29] Michele Boldrin and David K. Levine, *Against Intellectual Monopoly*, Cambridge University Press, 2008.

patents should be given only when the new drug had a different molecular structure and significantly greater therapeutic effect.[30] This was in a similar spirit to India's amended Patents Act of 2005.

Supplementing the new Patents Act of 1970 in helping indigenous drug manufacture were the Drug Price Control Orders and the Hathi Committee Report.[31] Their results were visible: the pharmaceutical market of the multinational drug companies in India came down from about 85 per cent before 1970 to less than 40 per cent by 1999.[32] India not only manufactured a significant part of its drug needs, especially of lifesaving drugs, it also went on to produce the API—which required a deeper industrial base—for most drugs. This transition in indigenous manufacturing was made possible by the scientific knowledge India had gained, and by people with the necessary industrial experience, as also by the CSIR laboratories that helped India develop alternate processes.

By the 1990s, India had emerged as an important global player in generic drugs, as well as API production. The Indian pharmaceutical industry is currently the largest global supplier of generic drugs—with an estimated 20 per cent share of the world's generic market.[33] This is exactly what Big Pharma had feared: that weakening the inherited colonial-era patents laws of most newly-independent countries would also weaken the control of Big

---

http://www.dklevine.com/papers/imbookfinalall.pdf

[30] Ibid.

[31] Report of the Committee on Drugs and Pharmaceutical Industry (Hathi Committee Report), 1975.
https://pharmaceuticals.gov.in/sites/default/files/Hathi_Committee_report_1975_0.pdf

[32] Sudip Chaudhuri, 'Public Enterprises and Private Purposes', *Economic and Political Weekly*, Vol. 29, 22, 28 May 1994.
https://www.epw.in/journal/1994/22/special-articles/public-enterprises-and-private-purposes.html
See also Sampath, 'Innovation in the Indian Pharmaceutical Industry'.

[33] Uday S. Racherla, 'Historical Evolution of India's Patent Regime and Its Impact on Innovation in The Indian Pharmaceutical Industry', in Kung-Chung Liu, Uday S. Racherla (eds.) *Innovation, Economic Development, and Intellectual Property in India and China*, 2019.

Pharma over the global market. Their fears were not limited to losing the market in the ex-colonies, but extended to the threat posed to their home markets.

Changes in the Patents Act were necessary for the Indian pharmaceutical industry to emerge, first in India and later at a global level. What is often forgotten in the story of the Indian drug industry is the contribution of the public sector undertakings: Indian Drugs and Pharmaceuticals Ltd. and Hindustan Antibiotics Ltd. Just as these two units had benefitted from the Haffkine Institute's experience in making vaccines, serum, and later, drugs, the Indian private sector also 'borrowed' people and knowledge from the public sector.[34] The public sector may be sick today, but we have to remember that our successes in the pharmaceutical sector owe much to its contributions.

The major contributions of P.M. Bhargava and his leadership of the Centre for Cellular & Molecular Biology (CCMB), Hyderabad, provided a firm basis to India's biologic revolution in medicines. It is thanks to the foundation laid by these Indian scientists that the country is today the largest manufacturer of generic drugs and vaccines in the world.

For many health activists in the 1980s, the key struggle was a rational drug therapy, creating an essential drug list, and regulating the prices of these essential drugs. The underlying belief, shared by many in the world, was that infectious disease had now been conquered and we had enough medicines in our kitty to control such diseases. The argument was that the latest patent-protected drugs could be substituted easily by older drugs in the essential list and were, therefore, not necessary for poor countries. The WHO had prepared an essential drug list, and certain health groups were

[34] An example is the founder of Dr Reddy's Lab. He got his Ph.D. from NCL, then worked for IDPL before starting out on his own. See also Manu Balachandran, 'Serum Institute: How an Indian Horse Breeder Built Asia's Largest Vaccine Company', *Quartz India*, 22 September 2015.
https://qz.com/india/506247/how-an-indian-horse-breeder-built-asias-largest-vaccine-company/

fighting to reduce the number of drugs and drug combinations in the market.

Those of us from the science and self-reliance movements agreed on the need to fight irrational drug combinations in the market; but we also believed that poor countries do need the latest in medicines. The medicine required should depend on a patient's disease, not their ability to pay, nor a country's wealth or poverty. The argument of restricting drugs to a small number of essential drugs made even less sense for India—the country already had a developing pharmaceutical industry with the capacity to manufacture a whole range of drugs.

It was the AIDS epidemic that brought together the two groups—of health and science activists. AIDS showed that the battle against infectious diseases was a continuous one; to think that the battle against such diseases was over was a foolish illusion. Disease is always going to strike, or strike back, and new drugs are a continuous requirement. This is what we have seen once again with the onslaught of SARS-CoV-2.

The AIDS epidemic also illustrated the utter heartlessness of Big Pharma, which was willing to sacrifice untold human lives in its hunger for profit. Big Pharma's price for AIDS drugs was $10,000–15,000 for a year's course—against India's price of $350. Big Pharma had set a price that 99 per cent of the 25 million AIDS patients at the time (over 38 million HIV+ individuals worldwide at the end of 2021) could not afford. The 'concessional' price offered was $4,000, twelve times more than the $350 at which Cipla was willing to sell the drug. Big Pharma did not stop there, and forty-one lawsuits charging violation of patent laws were filed against South Africa for its attempts to import generic AIDS drugs from India.[35]

---

[35] Jennifer Hillman, 'Drugs and Vaccines Are Coming—But to Whom', *Foreign Affairs*, 19 May 2020.
https://www.foreignaffairs.com/articles/world/2020-05-19/drugs-and-vaccines-are-coming-whom

The Untold Story of the Left in Indian Science

GETTING PAST THE TRIPS-WIRE: HEALTHCARE FOR THE PEOPLE

India had signed the World Trade Organisation/Trade-Related Aspects of Intellectual Property Rights (TRIPS) Agreement, which came into effect on 1 January 1995, but India had the benefit of a ten-year moratorium during which it was free to continue manufacturing drugs for its home market. Indian activists came together with global health activists to consider how to provide India's cheap generic drugs to countries in Africa that did not have an indigenous drug industry. They faced several questions: Under the TRIPS Agreement, could India still export drugs to Africa? What about other parts of the world? Or did the WTO trade rules bar such exports?

The battle was finally won in the Doha Round of the WTO negotiations in 2001, which reached the agreement that under conditions of an epidemic or a health emergency—and AIDS was held to be both—any country could issue a compulsory license. This would hold even for a company outside its borders, in this case an Indian company, to work its license. This is how the Indian generic industry became the supplier of 80 per cent of AIDS drugs in the world.

If the AIDS epidemic had exposed the weakness in the TRIPS Agreement when it came to dealing with public health emergencies, COVID-19 has brought out the underlying crisis of the health systems of the advanced countries as well. Why have countries with the most advanced health infrastructure and those with the strongest economies, failed to control the epidemic? The question becomes more pointed if we consider that a China, South Korea or Vietnam has managed better in controlling the epidemic.

In his book *Forgotten People, Forgotten Diseases,* the molecular biologist Peter Hotez writes about the two billion who face the threat of infectious diseases, people for whom Big Pharma is not interested in developing new drugs.[36] The last malaria drugs the US developed were for its soldiers in the war against Vietnam. The most frequently used TB drugs are now more than fifty years old. The question is: who has forgotten these diseases? If we add tuberculosis, malaria, dengue, and yellow fever to Hotez's list of forgotten diseases, the five billion people threatened by these infectious diseases certainly have not forgotten them.

Since the Third Plague pandemic (1890–1950) had killed relatively few people in rich countries, it encouraged the notion that only people in poor countries suffered from infectious diseases any longer.[37] All the advanced countries had to do was to keep such people—and their diseases—outside their borders. Like AIDS, the COVID-19 pandemic has once again proved that diseases will strike back: we are always just one mutation away from an emerging infectious disease.

This failure of the rich countries grows even starker if we consider their predictions from when the current pandemic began in China. Johns Hopkins and the Nuclear Threat Initiative came together to produce an index of which countries were best prepared to face the epidemic.[38] At the top of their list was the US, followed by the UK and other European countries. China, South Korea and Vietnam were all well behind. Needless to say, the index turned out to be pure fantasy.

Despite the HIV-AIDS epidemic, for people in the US the threat of a new infectious disease did not form a large part of their collective outlook. In the less affluent countries, where people continued to face infectious diseases—plague, cholera, small pox,

[36] Peter J. Hotez, *Forgotten People, Forgotten Diseases: The Neglected Tropical Diseases and Their Impact on Global Health and Development*, ASM Press; II edition, 2013.
[37] Figures indicate that about 1,700 died in Europe, against an estimated 15 million in Asia, with about 10 million in India alone.
[38] Global Health Security Index, Nuclear Threat Initiative and Johns Hopkins School of Public Health, October 2019.
https://www.ghsindex.org/wp-content/uploads/2020/04/2019-Global-Health-Security-Index.pdf

polio—they remained mindful of the public health measures needed during epidemics. Is this why the East and South East Asian countries, having recently faced SARS and the dangerous H5N1 avian flu, have fared better with COVID-19?

After the collapse of the socialist bloc, a triumphalist belief grew in the West—that the world could now be remoulded to suit the interests of global capital. Such a philosophy had many targets; one was the public health system, or what was seen as 'socialist medicine'. Privatising healthcare—including privatising publicly funded drug research—was the new paradigm pushed by the World Bank and the global think tanks that had mushroomed everywhere. This was the neoliberal phase of capital, which did not spare healthcare, municipal services, or other public monopolies such as electricity, telecom, and railways.

What happens when the hunger of capital enters the belly of the beast? In healthcare costs? And patented life-saving medicines? This is the question that the health activist from the left, Amit Sengupta, posed in 2018 at the last People's Health Assembly, held in Dhaka, Bangladesh. What happens when that raging hunger rides piggyback on a virus—and slips into those countries where infectious diseases are thought to belong to the remote past? Health activists had built on the global AIDS campaign and started an international health movement. The movement made the political choice to locate itself in the global South and not to be led by NGOs in the global North, however well-meaning they may be. Zafrullah Chowdhury, Amit Sengupta, and others realised that we cannot replicate the imperialist global order within the movement of resistance.[39] This is why the People's Health Assembly was founded in Dhaka in 2000. It led to the formation of the global People's Health Movement (PHM) and the Jana Swasthya Abhiyan (JSA) in India. The PHM has built a network of health activists who have kept the issue of public

---

[39] Zafrullah Chowdhury was the founder of Gonoshasthaya Kendra in Bangladesh, and a well-known public health activist.

health firmly on the agenda. They have been critical of the WHO's vulnerability to US pressure, or that of big private funders, and the drift towards more privatised healthcare—a development that would mean retreating from health as a public good. The WHO Watch, a team of volunteers from the People's Health Movement who attend the WHO bodies' meetings, follow debates, talk with delegates, and make statements to the Executive Board, was set up as a consequence of this larger engagement of health activists with the global agenda—which included issues from market-driven solutions of privatised healthcare to the battle over patents.

The engagement with the WHO made Amit Sengupta focus, once again, on intellectual property in the newly emerging area of biologics. Biologics comprise the cutting edge of new medicines for diseases from cancer to inflammatory diseases such as rheumatoid arthritis. Biologics have been making an entry even in antivirals. Biologics were priced very high, making it impossible for anyone in developing countries, except the super-rich, to access such drugs.

An example is Nexavar, a cancer drug which Bayer was selling at $65,000 for a year's course. India issued a compulsory license for Nexavar, instructing Natco to produce it in India. Marijn Dekkers, the CEO of Bayer, was quoted widely for calling this 'theft'. He was candid in explaining the basis of Bayer's price: 'We did not develop this medicine for Indians. . . . We developed it for Western patients who can afford it.'[40]

The COVID-19 pandemic has put public health and intellectual property rights back on the global agenda. Public health was important as long as infectious diseases were seen as threatening. Once this threat was 'forgotten', so was public health in rich countries. In the case of hospitals, what drove the system was private profit for private hospitals. The same capitalist criteria

---

[40] James Love, 'Bayer CEO Marijn Dekkers Explains: Nexavar Cancer Drug is for "Western Patients Who Can Afford It."' *Knowledge Ecology International*, 23 January 2014.
https://www.keionline.org/22401

were introduced as indices of 'efficiency' for public hospitals. The capitalist principle of maximising efficiency called for reducing beds, equipment, and medical staff; or, in the terms favoured by capital, 'rationalising' production and increasing 'efficiency'.

The other issue that will not go away is that of intellectual property rights—or patents. With the great and urgent demand for COVID-19 medicines and vaccines, it is very much back on the global agenda. The provisions for compulsory licensing—used as a tool by the developing countries—can be used during the COVID-19 pandemic. Any country can use the same compulsory licensing provisions to break the monopoly over a drug or a vaccine. This is the right given to every country by the Doha Declaration. The battle of people's health and medicines versus profit of capital is a continuing one, and COVID-19 is only another arena of this ongoing struggle.

9. Technology in a Postcolonial Setting: Notes from the Subcontinent

How do you transfer technology, especially as a postcolonial country? The subcontinent offers two distinct approaches: the aspiration of self-reliance in the case of India; and, as in Pakistan, the more 'quick fix' mode of transferring technology via import.

Engineers and technologists are generally an inarticulate lot and have written little about the nature of their craft. Narratives of technology from within the discipline are therefore rare while those from outside abound. There have been notable exceptions, including *Technology Acquisition in Pakistan* by Ghulam Kibria, written from the viewpoint of a practising technologist.[1] Subtitled the '*story of a failed privileged class and a successful working class*', Kibria's book is not just an indictment of the failure of the Pakistani elite, it also reveals the alternate path that was open to Pakistan: building on the work of *mistris*, craftsmen, whose role is generally lost to view in discussions of technology.

Kibria's book is a useful take-off point to discuss the role the state plays (and could play) in developing technology. It is fashionable today to write off the Nehruvian state as a total failure. The emerging policy prescriptions—often projected as models we should have followed after independence—endorse a hands-off state and free flow of capital, goods and technology. Nehru's insistence on developing the capital goods sector with state intervention was quite contrary to the views of various foreign experts, who wanted India to concentrate on manufacturing consumer goods. This was the prescription Pakistan too had received from US experts and faithfully followed.

---

[1] Ghulam Kibria, *Technology Acquisition in Pakistan: Story of a Failed Privileged Class and a Successful Working Class*, Karachi, City Press, 1998.

Technology in a Postcolonial Setting

The result, according to Kibria, was that Pakistan slid from being a state on the verge of take-off in 1947, to underdeveloped status. Kibria's insistence on the need for an interventionist state and his admiration for Nehru must have come as an unwelcome surprise to the economic establishment of the day. (While in today's India, a ruling party allergic to Nehru is far more likely to be displeased by his assessment than any economist.)

Underpinning Kibria's account of Pakistan's failure to acquire and develop technology is his view that no country or company ever transfers technology willingly. Technology is fundamentally about human beings acquiring knowledge and skills of a certain kind; skills that are then turned into designs, drawings and artefacts. This view of technology stands in contrast to the current economic thinking that technology is like any other commodity, open to trade. Kibria shares a little anecdote about this confusion among policy makers: Zia-ul-Haq appealed to the Toyota management to transfer automobile technology to Pakistan, to which they gently responded that Toyota would lose its market if they did so. Technology is intellectual capital, never to be parted with gratis. Even when a technology transfer occurs, it is always partial. It allows the companies to which technologies have been transferred to manufacture according to certain designs but never explains why these designs arose in the first place, thus precluding—or obstructing—future development. Countries that went on to indigenise technology after buying it have made research institutions a part of such technology transfers.

The second part of Kibria's book recounts the experience of Pakistan in developing technology, or rather in failing to do so. While India comes off better, particularly in the contrast between Nehru and the rulers of Pakistan, Kibria is fully aware of India's failures in developing a modern technological base, especially compared to the success stories of China and South Korea. He points out the importance not only of entrepreneurs in promoting innovation but also of state support. Back in the late 1790s, the US

Congress had supported Eli Whitney's project to mass-produce guns with interchangeable parts, and even awarded him a contract to supply the US army. Kibria links this precedent to the role later played by the state in South Korea and China to nurture technology development.

Not only did the state play an active role in promoting technology and innovation, it also helped protect the domestic market and defended the monopoly over any technology developed there. Britain had gone to the extent of banning the emigration of skilled workmen, routinely conducting searches and seizures to see that neither machinery nor designs left its shores.[2] The development of textile machinery in the US took place with Samuel Slater's secretive departure from Britain, carrying the entire design of the Belper spinning mill plant in his head. The plant at Belper in Derbyshire had been set up in 1776 by Jedidiah Strutt, using Arkwright's invention, the water frame for the spinning jenny. Slater was an apprentice in the Belper plant from the age of 14 till leaving for the US in 1789 at the age of 21. In England, he was reviled as 'Slater the traitor', while in the US he is regarded as the 'Father of American Manufactures'.[3] Even when Slater produced drawings of the spinning mill from memory, he had severe problems finding skilled workmen who could produce the required parts from these drawings. Eli Whitney had the same difficulty in finding skilled workmen, which drove him to set up one of the first modern mass production factories for engineering goods.[4] Though Eli Whitney is widely credited as the father of

---

[2] Britain passed a series of laws banning the emigration of skilled workmen/artisans and machinery for a range of industries including textiles. David J. Jeremy, 'British Textile Technology Transmission to the United States: The Philadelphia Region Experience, 1770–1820', *The Business History Review*, Vol. 47, 1, pp. 24–52, 1973.
https://doi.org/10.2307/3113602

[3] Neil Heath, 'Samuel Slater: American Hero or British Traitor?' BBC News, 22 September 2011.
https://www.bbc.com/news/uk-england-derbyshire-15002318

[4] Though Eli Whitney is widely known as the father of mass production—in

mass production, the idea predates him. The invention that did most to make him famous was the cotton gin, mechanically separating the seeds from the cotton fibre. Instead of reducing slavery, it only intensified the use of slave labour, which goes to show that technology improvements do not in themselves lead to social change.

The invention that made possible mass production and interchangeable parts is the milling machine. Whitney was not the first to produce a milling machine. Prior to its introduction, a skilled workman would make every part of the equipment or machine and put it together on his own. With the milling machine, factories that produced guns broke down this process so that every workman produced only one part. As the parts were interchangeable, they could be assembled together by other workmen, considerably speeding up the process of manufacture. The key to interchangeable parts was the milling machine. This was also the first step in the eventual de-skilling of the worker: with advances in automated manufacture, s/he was transformed from a skilled craftsman to a worker with ever fewer skills. The 'skills' were transferred to the machine by breaking up the processes in such a way that machines did the same tasks. They were first propelled by water, then by steam engines.

The historical material that Ghulam Kibria has put together in his analysis of Pakistan allows him to comprehend the process of technology acquisition. It is clear to him (though not to a number of policy planners in Pakistan and India) that technology transfers are never gifts from one nation to another. Historically, such transfers have taken place when people with technical expertise migrated to new countries, or technologists/workers learned skills abroad and returned home. The conundrum of technology transfer is that

American manufacturing—and of the milling machine, there were others who may have played a bigger role but could not match his public relations. Robert S. Woodbury, 'The Legend of Eli Whitney and Interchangeable Parts', *Technology and Culture*, Vol. 1, 3, pp. 235–253, 1960. https://doi.org/10.2307/3101392

a 'donor' transfers technology only when the recipient already has the capability to develop it himself. The choice before donors then is either to face a new competitor or retain the other as an ally through appropriate licensing. Kibria identifies the support—different forms of state intervention—required for the success of technology acquisition.

Kibria's account of Pakistan's technological abilities before independence is fascinating as such a history needs to be written for India too. He chronicles the three areas in which Pakistan had developed international quality skills. At the time of independence, Pakistan had world-class skills in the design of infrastructure—railways, canals, bridges, and roads. The Pakistani *mistris* had developed fairly advanced oil engines and machine tools, using their knowledge and by reverse-engineering old machine tools discarded by the Mughalpura Railway Workshop. It is interesting that Kibria locates the development of Pakistan's technological capabilities in the development of railways—the first modern technology that entered the Indian subcontinent to meet British colonial needs. Marx had already pointed out the importance of the railways to India's technological development; the history of technology clearly substantiates Marx's remarkably prescient observation.

Kibria suggests that Pakistan's ability to design infrastructure, particularly railways, canals, roads and bridges, could have been leveraged internationally, particularly in other Third World countries. This would have led to further development of these skills and brought in valuable foreign exchange. Instead, Pakistan let these skills erode and today has to invite foreign consultants to carry out the same tasks. Perhaps Kibria does not sufficiently take into account that quite often infrastructure was developed in Third World countries with foreign aid, and this was tied to the import of consultancy and equipment from donor countries. In such a context, even if Pakistan's skills were world-class, breaking into the international market would not have been that simple.

While infrastructure developments in the subcontinent were largely driven by British colonial needs, indigenous developments in oil engines and machine tools had quite different origins. The Mughalpura Railway Workshop had 10,000 skilled workers, a critical mass for developing new technologies. Using machines that had been disposed of by the workshop, some of the skilled workers who had retired set up companies manufacturing oil engines and simple machine tools. The Pakistani (and Indian) *mistris* were, and are, famous for *jugad* (improvisation). This was the creative source of their ability to manufacture engines and machine tools at much lower prices than similar imported equipment. Kibria traces the origin of the famous Batala lathe (associated with Batala, Punjab) to the Lahore machine tools industry.[5] Not only were engines and machines tools developed using reverse engineering and the ingenious *jugad* of the *mistris,* but they also came to be widely used by the British during the Second World War and were subsequently exported. However, instead of leveraging this legacy, Pakistan decided to bank exclusively on imported technology for its development. The government followed the advice of the Harvard Advisory Group, who viewed Pakistan as a country with no technology whatsoever.[6] Even though the machine tools segment—a vital component of the capital goods sector—was already developed in and around Lahore, these economists advised that Pakistan should concentrate only on the consumer goods sector. Unfortunately, the political and bureaucratic leadership of Pakistan—and later, its military rulers—also shared this view, with the result that its technological legacy was squandered.

Kibria is also unsparing in his criticism of Pakistan's business

---

[5] As an engineer working in a factory in Howrah, I learnt how important the Batala lathe is to the entire ancillary industry that had grown up around the engineering industry in West Bengal. Its history definitely needs to be written.

[6] S. Akbar Zaidi, 'Special Report: The Changing of the Guard 1958–1969', *Dawn*, 2 September 2017. https://www.dawn.com/news/1355171

class. The business elite was more interested in cosy crony capitalism and collecting various forms of largesse from the state. Kibria traces the problem back to this lack of entrepreneurship and a servile attitude to the Pakistani political elite. While the working class had the innovative flair to develop technology, the feudal elite failed to take this forward due to their lack of entrepreneurship.

Kibria's account of Pakistani failures is important for policy planners on this side of the border as well, especially now that we seem embarked on a similar journey to the one Pakistan began fifty years before us.[7]

---

[7] Having formally given up self-reliance in the 1990s, and the Planning Commission in 2014, we now travel the same path as Pakistan did after independence. Instead of the Harvard Advisory Group, we have as advisors a number of management consultants such as McKinsey, Boston Consulting, Deloitte, PWC, and KPMG. Their advice is always the same—Liberalise: allow foreign capital, decrease import duties and provide more protection to foreign capital.

Section IV

Planning the Republic of Reason

10. Hindutva, Pythagoras and the Zero

The BJP's attacks on reason and critical thinking, on scientific temper, are an integral part of its assault on the Indian nation. Article 51 A (h) of the Constitution demands as a part of fundamental duties that citizens 'develop the scientific temper, humanism and the spirit of inquiry and reform'. In October 2015, more than 100 leading scientists of the country felt compelled to point out that ' . . . what we are witnessing instead, is the active promotion of irrational and sectarian thought by important functionaries of the government'.[1] This attack on reason and science is also an attack on history, replacing the actual advances in Indian mathematics, astronomy, and medicine with myths. In positing mythology as history, it destroys the basis of how to study history using reason and scientific tools. Instead, science and history become the playground of a perverted nationalism. By negating the actual advances made in India and replacing them with myths, it also destroys the basis of future advances in which critical reason and a scientific outlook have to be a part of any discipline. We will examine some of these vainglorious claims about India's past achievements and contrast them with a few of India's actual achievements.

THE BATRA SCHOOL'S HISTORIOGRAPHY OF SCIENCE

RSS ideologues take the imagination of mythology and present it as a matter of fact, of history. When it comes to the scientific

---

[1] Sachin Kalbag, 'More Than 100 Scientists Speak Out Against Bigotry', *The Hindu*, 28 October 2015.
https://www.thehindu.com/news/national/more-than-100-scientists-speak-out-against-bigotry/article7813958.ece

imagination—how to generate new advances—they fall back on the sterile claim that we have nothing new to discover: it has all been done already by our sages. This is the version of science, as mumbo-jumbo, being championed by figures like Dinanath Batra, a former general secretary of the RSS's school network, Vidya Bharati. Batra also runs the RSS offshoot, Shiksha Bachao Andolan Samiti. In *Tejomay Bharat*, a series of books authored by him—with a foreword from Narendra Modi—and introduced in 2014 as supplementary readers on the Gujarat state-school syllabus, Batra makes the following claim: 'America wants to take the credit for the invention of stem cell research, but the truth is that India's Dr Balkrishna Ganpat Matapurkar has already got a patent for regenerating body parts . . . . . You would be surprised to know that this research is not new and that Dr Matapurkar was inspired by the Mahabharata'. There follows proof positive of stem cell research in ancient India: a mass of flesh ejected by Gandhari's body was stored with ghee in a hundred vats, from which, at the end of two years, emerged the hundred Kauravas. If Sanjay can describe the Mahabharata War to Dhritarashtra as they sit far from the battlefield, it demonstrates the existence of iron-age television. The graft of an elephant head on Ganesh was an instance of cosmetic surgery!

In his October 2014 speech at the opening of a new wing of the Reliance Foundation's hospital in Mumbai, Modi endorsed Batra's claim that genetics and organ transplants were available in ancient India. 'What I mean to say', proceeded Modi, 'is that we are the country which had these capabilities. We need to regain them'.

Not only did we in the past have all the knowledge we needed, we also gave it to others. The Hindutva brigade have been proclaiming that the Greeks learnt Pythagoras' theorem from India. This is what Harsh Vardhan, the BJP government's Minister of Science and Technology from 2014 to 2021, claimed at the Indian Science Congress in 2015: that 'we', having discovered the Pythagoras theorem, gave it 'magnanimously' to the Greeks.

Absurd claims have also been made for Vedic mathematics, which is being introduced into the school curriculum of Gujarat and is one of the projects being promoted under the National Education Policy of 2022.

The Rashtriya Sanskrit Sansthan, under the Ministry of Human Resources, declared on its home page, 'Sanskrit . . . . provides the theoretical foundation of ancient sciences'. No need to study science, just study Sanskrit. Development is not necessary, regaining 'ancient' knowledge will do. No wonder the science budget, as a percentage of the national budget, has been continuously cut under the Modi regime.

What the RSS and the Hindutva forces propagate is 'belief' in place of evidence and reason. Based on this belief, myth and fantasy become the real past. Only the Vedas, Ramayana and Mahabharata are to be used as evidence; all other evidence—of linguistics, archaeology and texts—can be safely disregarded. If scientific evidence—say, the carbon dating of artefacts—shows that the Vedic Age is 3,500 years old, that evidence is clearly no good, since the oral tradition claims for it a past of 10,000 or 20,000 years. Similarly, a Pushpak Vimana or a Brahmastra are to be believed as actual flying machines and nuclear weapons known to the 'ancients'.

All that scientists need do—according to Batra and his fellow travellers—is learn Sanskrit and rediscover ancient science. Presumably, they need not bother about learning science in school and college, or go through the drill of developing theories and performing actual experiments for their verification. In the BJP's scheme of things, Sanskrit departments are competent authorities to pronounce on everything, from science to ancient Indian history. It was the Sanskrit department at Mumbai University that organised a special session to accompany the Indian Science Congress in 2015: its findings were that the Rig Veda is 5,000 to 10,000 years old, that the Aryans were no migrants into India, and that we had flying machines.

A scientific understanding of history is an exercise of carefully sifting through the evidence, building a coherent narrative of how people lived and how society developed. Texts—written or oral— must be validated by other evidence. History is as much a subject of scientific inquiry as any other discipline. It is not a glorification of the past but a critical examination of all developments—negative as well as positive. This is anathema to the Dinanath Batra school of historiography. To their thinking—or lack of it—only self-flattery counts. Reason has no purpose unless it endorses belief. And belief is reserved for the mythical glory of a Hindu past.

Such a view in which myth masquerades as reality is not only damaging to history, but also to science and mathematics. This mode of fabricating knowledge is not new—it was practised extensively in India and Europe in the past. It meant learning by rote and reading only canonical texts, while relegating all experiments and examination of nature as the task of the 'lower' classes; or, in India, 'lower' castes. Europe's monasteries privileged ancient knowledge over what developed in the living practice of societies. Thus, a physician's studies in medieval Europe involved three years of logic, and then one or two years of ancient texts. All of it by rote and no knowledge of everyday practice. This was also the method of study that destroyed all knowledge of India's past in the so-called centres of Indian learning. The Gurukuls of Benares, on being shown the Ashokan pillar inscriptions from Sarnath, had no knowledge of what they meant; their learned texts had no information on Ashoka whose edicts were turning up all over India. The Brahmanical texts had destroyed all knowledge of Buddhism in India, and therefore in their books Ashoka's reign never existed!

INDIAN MEDICINE, SURGERY AND MATHEMATICS

Such a self-serving, vainglorious view of history misses all the real advances that India made, in science and mathematics as much as other fields, where practitioners of knowledge made significant advances that receive little attention from religious zealots today. For instance, Indian medicine shows a decisive break with the use of rituals and practical medicine prescribed by the Atharvaveda, for an evidence-based system shown in the *Charaka Samhita* and *Susruta Samhita* from the early centuries of the Common Era.[2] The class-caste features of the *Charaka Samhita* exist, but it is still a departure from the Atharvaveda, which is mainly incantations against evil spirits causing disease. Charaka uses empirical knowledge in the practice of medicine. Susruta develops surgery, which means handling blood, pus, and other 'polluting' substances. With Ayurveda, medicine gets converted to a Veda, or received wisdom. Ayurveda should not be confused with the folk wisdom of communities who use empirical local knowledge and common herb lore. It became elite medicine, practised by Brahmins, and lost its empirical character.

Susruta had developed detailed surgical procedures, including cosmetic surgery—real cosmetic surgery which included reconstruction of the nose and ears, unlike Modi's example of the elephant head and Ganesha. Cosmetic surgery survived in India, and was reported by the British. Interestingly, it was performed by a Vaid of the potter caste in Maharashtra.[3] This is similar to the West, where, unlike the doctors who were schooled in ancient texts and logic, it was the barbers who became surgeons; in England, they belonged to the Royal College of Barbers and Surgeons.[4] The

---

[2] *Charaka Samhita* moves away from 'magicoreligious therapeutics to rational therapeutics' and the use of diet and drugs for 'directly perceptible results.' Debiprasad Chattopadhyaya, *Science and Philosophy in Ancient India,* Aakar Books, 2013.

[3] Two English doctors had witnessed nose reconstruction surgery in Pune, done by a potter surgeon, and wrote it up for the *Madras Gazette*, an article reproduced in *Gentleman's Gazette* in 1794. The procedure is detailed by the *Sushruta Samhita*, but was not a part of later Ayurvedic practice.

[4] The Company of Barber-Surgeons, established in 1540, was a trade guild and a London Livery Company that apprenticed and examined trainees within the City of London, *The History of the RCS* (Royal College of Surgeons).

progress of human knowledge is not a command performance by history to suit the preferences of today's rulers, it moves by a different logic.

The Pythagoras theorem: In its most well-known version, the Pythagoras theorem states that in a right-angled triangle, the square on the hypotenuse (the side opposite the right angle) is equal to the square on the other two sides. A set of numbers that satisfies this relationship is called Pythagorean tuples.

Hindutva's claim to a patent on the Pythagoras theorem is based on the Sulba-sutras, texts that tell us how to make different kinds of altars or vedis for religious purposes. There are four important Sulba-sutras, that of Baudhayana (c 800 BCE), of Manava (c 750 BCE), Apastamba (c 600 BCE), and Katyayana (c 200 BCE). The Sulba-sutras are a part of Vedanga Jyotisha, and are therefore a part of Brahmanical rituals.

The Sulba-sutras give us formulations of the theorem, not as we express it today but using rectangles instead of triangles. It is clear from the Sulba-sutras that the authors knew the geometrical relationships between the three sides of a right-angled triangle and also knew a set of numbers that satisfy this relationship. This allowed for the construction of complicated fire altars by being able to create right-angled triangles.

The Pythagoras theorem was known in India, Babylon, China and Egypt, and was used for land measurement, the construction of religious structures (the Egyptian pyramids, fire altars) and for constructing canals (Babylon).[5] There are Babylonian tablets from 1800 BCE, considerably predating the Sulba-sutras. Used for teaching scribes, the tablets show a working knowledge of

---

https://www.rcseng.ac.uk/about-the-rcs/history-of-the-rcs/#:~:text=The%20Company%20of%20Barber%2DSurgeons,to%20establish%20their%20own%20identity

[5] Rahul Roy, 'Babylonian Pythagoras' Theorem, the Early History of Zero and a Polemic on the Study of the History of Science', *Resonance*, Vol. 8, January 2003.
https://www.ias.ac.in/article/fulltext/reso/008/01/0030-0040

Hindutva, Pythagoras and the Zero

Pythagorean tuples and the Pythagoras theorem. It is also known that Babylon and Egypt routinely exchanged goods, knowledge and texts. Whether independently or through its trade with Babylon, Egypt also knew about the Pythagoras theorem. It is known that Pythagoras (c 570–c 490 BCE) spent a considerable part of his early life in Egypt and learned his mathematics from the Egyptians.[6] By the sixth century BCE, possibly earlier, the Chinese knew about the Pythagoras theorem, or the kou-ku/go-gu (width and length) theorem as it later came to be called.[7] Pythagoras' claim to fame here is that he was supposedly the first to provide formal proof of the theorem. Baudhayana provides a proof for right-angled isosceles triangles, while Apastamba gives a numerical proof of the more general statement, true for any right-angled triangle. The Chinese, too, had developed a proof of the Pythagoras theorem, but in a visual form.

Neither is the Eurocentric claim valid that Pythagoras provided the first proof, nor the Indocentric view that India was the originator of the theorem.

The Pythagoras theorem shows how the history of science and mathematics is not one of who did what first, but of appreciating the broad sweep of development and the contributions of each cultural area, an exercise rather different from the competing claims of cultural supremacists.

Zero and the Indian number system: We need to appreciate on similar lines the history of India's development of the modern number system and the zero. Apart from the actual symbols of the numbers, there are three important elements that constitute the

---

[6] Pythagoras was from Samos, an island off the coast of modern Turkey and was an Ionian Greek from Anatolia. Whether he knew about the truth of the properties of the right-angled triangle or gave a proof of it is disputed, as is much of his legacy. There is no evidence to show that Pythagoras had provided the geometrical proof that later Greeks attributed to him, but it is possible he knew of the relationship between the three sides as did many in other civilisations.

[7] G.G. Joseph, *The Crest of the Peacock*, Princeton, NJ: Princeton University Press, 2000.

modern number system—place-value notation, base 10, and the use of zero as a number.

Babylon and India made major contributions in the development of place-value notation. In this notation, a digit (0–9) has a value based on its position in a number; e.g., in 66, though both the numbers are sixes, the value of the first one is sixty, and of the second six. In Roman numerals, it would be written as LXVI, where L is fifty, X is ten, V is five and I is one.

The Roman number system makes computation extremely cumbersome. Place-value notation makes the basic arithmetic operations of addition, subtraction, multiplication and division much simpler, allowing for the complex calculations required by commerce, engineering, astronomy and physics. Claudius Ptolemy, in his astronomical text *Almagest* (second century CE), uses the Babylonian system for his calculations, though he expresses them finally in Roman numerals.[8] Though the Babylonians had developed place-value notation earlier, they used a sexagesimal system, or base 60, while India used the base 10 (decimal) system, which we all now use. The Babylonian sexagesimal system persists in our calculation of time: an hour is divided into 60 minutes, and one minute into 60 seconds.

Aryabhata marks the beginning of the classical age of Indian mathematics. Born in 476 CE, he formulated place-value notation in his magnum opus, *Aryabhatiya*. He enunciated the rule '*Sthanam sthanam dasa gunam*', meaning from place to place, multiply by 10. The Bhakhshali manuscript, a set of seventy leaves of birch bark found near Peshawar, is composed of material from at least three different periods. The parts dated to the third or fourth century CE contain the earliest written record we have of the *modern symbol of zero* and its use in place-value notation.[9]

[8] The title, like its earliest surviving manuscript, is of later provenance and derives from the Arabic-Greek hybrid of *al-magisti*, illustrating the multicultural journey through which knowledge evolves and gets preserved.
[9] Bodleian Libraries, 'Carbon Dating Finds Bakhshali Manuscript Contains Oldest Recorded Origins of the Symbol "Zero"', *Gardens, Libraries and*

Hindutva, Pythagoras and the Zero 219

The use of a symbol to mark the absence of a number is again available from Babylon. Babylonians used various symbols to indicate the lack of a number in that position in the place-value system. Ptolemy, also, used a null symbol in his calculations.

The major contribution of Indian mathematicians was to treat zero not as a null symbol, but as a number, and provide the rules for its use in arithmetical operations. Varahamihira, a younger contemporary of Aryabhata in the sixth century CE, was the first to use zero in mathematical operations. To see how knowledge was regarded as international, we may note that in Varahamihira's compilation of astronomical knowledge, *Panchasiddhantika* (the five canons), he attributes two of them to foreign sources, Paulisa (of Greek origin, probably Alexandria) and Romaka (of Roman or Greek origin). Brahmagupta, in the seventh century, was the first to formulate the mathematical rules for using zero as a number and also the concept of negative numbers.[10] While he correctly formulated rules for addition, subtraction and multiplication by zero, he ran into problems with division by zero. Modern mathematics has 'solved' this problem by banning the operation itself!

It is also interesting to note that the Indian number system not only evolved the three major components of the system— place-value notation, base 10, and use of zero as a number—it also provided the current symbols of the numbers themselves. These are derived from numerals used in the Brahmi script. Brahmi was widely used to write Prakrit, in which the major Jain mathematical texts were composed. Unlike the Vedas or the Vedanga Jyotisha, Jain mathematical texts were not based on Brahmanical rituals. Jains believed, much like the Pythagoreans, that mathematics itself was an exercise in reaching higher consciousness. In their endeavour to reach infinity the Jains also developed notation for very large numbers.

Museums, Oxford, 14 September 2017.
[10] Rahul Roy, 'Babylonian Pythagoras' Theorem'.

VEDIC MATHEMATICS VERSUS REAL MATHEMATICS

As always with the Hindutva brigade, the real advances that took place in Indian mathematics do not satisfy them. They must also make fraudulent claims for the wisdom of the Vedas and its advanced knowledge. Such claims were made by Swami Bharati Krishna Tirtha, the late Shankaracharya of Puri, regarding what he called 'Vedic' mathematics. Professor S.G. Dani and others have shown that this has nothing to do with ancient mathematics or the Vedas.[11] Though the antiquity of the Atharvaveda is claimed for it, no such mathematical evidence has been found in the existing literature on the Vedas.

This so-called Vedic mathematics is just a set of tricks or gimmicks for certain limited arithmetical calculations. It has little to do with mathematics as we know it today, and has virtually no value in the age of calculators and computers.

The other problem with the Hindutva brigade's focus on the Vedas and the epics—Ramayana and Mahabharata—as the sole source of wisdom is that it misses out on all the advances Indian mathematics made in a later period. Aryabhata had derived the value of pi to 5 decimal places, the most accurate derivation of his period. Indian mathematicians such as Aryabhata, Brahmagupta, Bhaskara I (fifth to seventh century CE), and later, Bhaskara II (twelfth century), made significant contributions to trigonometric functions and the construction of trigonometric tables. The work of Indian mathematicians such as Madhava, Nilkantha and others from what is now termed as the Kerala School (of the fourteenth to sixteenth centuries), gave us the series expansion of trigonometric

---

[11] S.G. Dani 'Myths and Reality: On "Vedic Mathematics"', *Frontline,* updated from his two-part article from 22 October and 5 November 1993.
https://www.tifr.res.in/~vahia/dani-vmsm.pdf
The full list of his writings on this subject, including '"Vedic Mathematics": A Dubious Pursuit', is available online.
http://www.math.tifr.res.in/~dani/

Hindutva, Pythagoras and the Zero

and other functions that predate work on this in Europe.[12]

The form of mathematics as we know it today—from axiom and hypothesis to proof—was first given by Euclid of Alexandria (fl. 300 BCE) in his textbook on geometry, *Elements*. This was very much a part of the Greek development of mathematics. The Greeks also introduced proof using *reductio ad absurdum*: an assumption that leads to a logical contradiction must be the wrong assumption. This is a powerful tool in the armoury of mathematics. Other civilisations, such as Babylon and India, used far superior computation systems to that of the Greeks, but did not develop this formal method of proof.

It is very much a Eurocentric view that condemns all other civilisations that did not develop the formal methods of the Greeks as 'algorithmic', confined to 'practical' mathematics, and therefore inferior. What such a view overlooks is that trade, commerce, and developments in science and technology rest precisely on this 'practical' ability to compute with numbers.

This modern number system was developed in Babylon and India, and then transmitted to the Arabs. Al-Khwarizmi (780–850 CE) wrote a treatise on the Indian system of numerals—*Kitab al-jam' wa'l-tafriq bi-hisab al-Hindi*, which survived in a Latin translation as *Algoritmi de numero Indorum* (al-Khwarizmi on the Hindu art of reckoning). The word algorithm is derived from the Latin corruption of his name, just as the *Almagest* derives from an Arabic corruption of Greek. Leonardo of Pisa, more commonly known as Fibonacci, popularised the Indian system of numbers in Europe through his book, *Liber Abaci* (book of calculations), 1202 CE. Without such developments, we would not have had the industrial revolution or the birth of modern science in Europe.

There was also an exchange between all these cultures, particularly between India, Greece, Babylon and Egypt. There is a story about the great eleventh-century polymath, Ibn Sina (known

[12] Kim Plofker, *Mathematics in India*, Princeton, NJ: Princeton University Press, 2009.

in the West as Avicenna), which relates that his father told him not to waste his time but go and learn some mathematics from the Hindu merchant in the market.

Different cultures developed knowledge about geometry and calculations with numbers, to different degrees of sophistication. If the formal nature of Greek mathematics has been a powerful instrument in the development of modern mathematics, the growth of computational methods and algorithms are the bedrock of modern science and technology.

The one unfinished matter that I have not dealt with here is that if India had made major advances in mathematics and the sciences in the past, why did it fail to develop further? This is the same question that Benjamin Farrington had posed in his analysis of Greek and Roman science. His answer: it was the separation of the hand and the brain.[13] If thinking and doing are separated—even worse, if work is subordinated to thought—both thought and work lose their creative impulse. It is in the doing that we interact with nature and learn what it really is. And, of course, artefacts as instruments of enquiry, from the telescope to the microscope, allow us to advance knowledge.

Indian science suffered far worse by separating thought and labour.[14] Even writing was subordinated to memorising as a method of learning. Instruments were the territory of the 'lower' castes, therefore 'higher' knowledge was without instruments, reached by 'pure' thought. The only knowledge system that could still grow under such conditions, where all labour was held to be polluting, was mathematics. Though mathematical texts exist, most of the work was done using numbers written in dust, or dhuli-karma.

---

[13] Benjamin Farrington, *Head and Hand in Ancient Greece*, London, Watts, 1947.

[14] Chattopadhyaya points out that while Vedic composers compared their compositions to those of craftsmen and the wisdom of action, by the time of Upanishads, manual labour and all those who perform it had been relegated to lower orders. Debiprasad Chattopadhyaya, *Science and Philosophy in Ancient India*.

Hindutva, Pythagoras and the Zero

Most texts were more of an aid to memory, as wisdom was passed by word of mouth and through rote learning in the guru-shishya tradition. Under such conditions, mathematics is the only branch of knowledge in which the subcontinent made major advances, from the number system to the developments of the Kerala School.

The caste system imposed a far stronger division of knowledge than in other countries where the aristocracy also looked down on manual labour. Even in Europe, the division between landowning aristocrats and the tradespeople who sullied their hands with labour continued for a long time. In Great Britain, where the landowning aristocracy looked down on all other sections, engineers were not classified as 'gentlemen' and stood much lower down in the pecking order from scientists, mathematicians and lawyers.[15] The Hindutva lobby attempts to invert the arguments of the Eurocentric or Orientalist account of mathematics, using mostly empty assertions and fraudulent history. The answer to Eurocentric 'history' is not substituting it with fraudulent history, but with the real history of knowledge. And such a history does not support supremacist claims, whether of the Eurocentric or Hindutva kind. Confusing history with fantasy also ignores the central division that caused the ossification of Indian science, the separation of the hand from the head.

---

[15] In Victorian times, engineers were not recognised as 'gentlemen', unlike army officers or the members of the clergy. David Cody, *The Gentleman*, Victorian Web.
https://victorianweb.org/history/gentleman.html

11. Building a Nation with a Scientific Vision

NATIONAL LIBERATION STRUGGLES AND CONTESTING VERSIONS OF THE NATION

National liberation struggles in different parts of the world have found that the 'blood and soil' framework of European nationalism—or ethnic nationalism—does not work for them. Most of the colonised countries were composites of multiple identities—religious, ethnic, and linguistic—owing partly to the arbitrary carving-up of conquered territories by the colonial powers.[1] The colonial powers also created divisions in the colonies on the lines of religion, language, and ethnicity. In contrast to the divide-and-rule imperial game, the task of national liberation struggles was to unite the people in spite of these multiple identities.

There is an interesting contrast in the way international—that is, Western—media treats social divisions in Europe and in Africa/Asia. In Europe, ethnic groups such as the Basques, Serbs, Croats, and others are not called tribal, but any such division in the ex-colonies is always termed as tribal, not ethnic, reinforcing the idea that, apart from the West, all others are backward societies still fighting tribal wars. Never mind that the 'tribal' divisions between the Germans and the French in Europe have imposed two world wars on us!

The left, along with other leaders of independence movements against colonialism, located the nation on the terrain of an

---

[1] The Berlin Conference, 1884–85, is held to be when and where the partition of Africa into colonial territories was formalised, but the process had begun earlier and would continue till later with bilateral agreements and changes. The partitioning of Africa, not by people, but executed clinically by the colonial powers, is visible today. Unlike boundaries based on where people live, these borders are straight lines across the map.

Building a Nation with a Scientific Vision

independent national economy.[2] The key element in all such national struggles was securing control of the state apparatus through which imperial oppression—economic and physical—had been carried out. The starting point of the struggle against imperialism was the struggle against the colonial state, a struggle that united the people and shaped the anti-colonial, national consciousness. These two elements—a secular basis for uniting the people, and economic development—distinguished the vision and action of all the leaders of independence movements in various parts of the world; including, of course, in India. In the language of writers on nationalism, this was civic nationalism, investing all the inhabitants with rights by virtue of their citizenship in the nation.

The nation was not part of the constitutive slogans of the French revolution: the terms around which the revolutionary upsurge had revolved were, liberty, equality, and fraternity. However, the first challenge before this new state was organising its defence against the neighbouring countries which had declared war on France, intending to crush the revolutionary republican order. If earlier wars for the kingdom had been fought in the name of the king, in whose name would the French army now fight? The idea of France as a fatherland, for which its citizens should fight, was created to meet this challenge. The need to unite people in defending their 'fatherland' against invasion gave rise to the 'modern' European nation state; nationalism was the fortifying ideology of this nascent bourgeois state.

SLAVERY, GENOCIDE AND LOOT: FROM CHRISTIAN CIVILISING MISSION TO WHITE MAN'S BURDEN

European nationalism, including French nationalism, had no difficulty in combining two contradictory ideas: that of citizenship

---

[2] Manu Goswami, *Producing India: From Colonial Economy to National Space*, University of Chicago Press, 2004.

for people within the home country, and debarring from citizenship the indigenous dwellers of the colonies.[3] One of the key elements of the new national consciousness was the perceived superiority of the people constituting 'our nation', a belief that would rapidly convert civic nationalism to an imperialist ideology.

European 'explorers' such as Christopher Columbus and Vasco da Gama had 'discovered' lands already inhabited, and established the first European colonies. The Spanish and Portuguese enslavement of people in Africa and the Americas was justified by the *Doctrine of Discovery*, a bull issued by Pope Alexander VI in 1493,[4] which essentially argued that non-Christians were not fully human and could therefore be dispossessed of their lands, enslaved, or even killed by the Christian colonisers.[5]

The European colonisers' belief in the superiority of Christians as a people transformed itself painlessly into the ideology of 'scientific racism'.

The enlightened white man was tasked by destiny with bringing 'civilisation' to the natives.[6] It was his racial obligation, this White Man's Burden.[7] Both 'civilising missions'—the earlier Christian one, and the later, explicitly racist one—were accompanied by slavery, genocide, loot, and plunder. Portugal and Spain had taken

[3] C.L.R. James, *The Black Jacobins: Toussaint L'Ouverture and the San Domingo Revolution*, Vintage Books, 1989.
[4] Excerpt from the *Doctrine of Discovery*, Upstander Project. https://static1.squarespace.com/static/54f8b4cfe4b0b230c7abfe97/t/56f162baf699bb3c4bac74fd/1458660026512/Doctrine+Inter+Caetera+EXCERPT+March2016.pdf
[5] Roxanne Dunbar-Ortiz, *"Not A Nation of Immigrants": Settler Colonialism, White Supremacy, and a History of Erasure and Exclusion*, Beacon Press, 2021.
[6] Aubrey Clayton, 'How Eugenics Shaped Statistics', *Nautilus*, 27 October 2020. Eugenics and statistics were closely connected to racial politics in the US, UK and Germany, as can be seen in Stefan Kühl, *The Nazi Connection: Eugenics, American Racism, and German National Socialism*, New York and Oxford: Blackwell Publishing, 2002.
https://nautil.us/how-eugenics-shaped-statistics-9365/
[7] As expressed in Kipling's poem of the same title, published in 1899, which exhorted the USA to annex the Philippine Islands for the good of its ('half devil and half child') natives.

Building a Nation with a Scientific Vision

an early lead in colonising Africa and the Americas, but all the major European nations participated in the slave trade and in using slave labour. The English and the French built extensive sugar plantations in the West Indies based on 'modern' slavery. It is instructive to see how the French revolution's emancipatory vision of the citizen—every person residing within the geographic boundary of France—collapsed when it came to the colonies. In Haiti, then a French territory, the revolutionary government of France was faced with a revolt by slaves who raised the slogans of the French Revolution—liberty, equality, and fraternity.[8] The government in Paris dithered for a while but ultimately denied freedom to the slaves. The colonial subjects were not citizens after all.

It was the slave trade from Africa and colonial plunder from India and other parts of the world that 'financed', or provided the necessary capital for the industrial revolution. As Marx noted, 'capital comes dripping from head to foot, from every pore, with blood and dirt'.[9] It also led to the destruction of the weaving community and to de-industrialisation in India. Marx writes vividly about this: 'The misery hardly finds a parallel in the history of commerce. The bones of the cotton-weavers are bleaching the plains of India.'

TECHNOLOGY AND NATIONS

The story of 'normative' European nationalism derives from a handful of countries in Western Europe—France, England, Germany. Other European empires and kingdoms, whether we consider Tsarist Russia, the Ottoman Empire, or the Austro-Hungarian Empire, were home to many languages and also diverse in their confessional and ethnic composition, besides being spread

---

[8] C.L.R. James, *The Black Jacobins*.
[9] Karl Marx, *Capital*, Vol. I, Chapter 31.
https://www.marxists.org/archive/marx/works/1867-c1/ch31.htm

across a vast swathe of Europe. The concept of nationalism derived from the narrow context of English and French pre-imperial history is unrepresentative of Europe as a whole, and still less relevant as a model for emerging patterns of nationalism in Asia, Africa, and Latin America.

Benedict Anderson identifies the rise of the nation—an imagined community—with the rise of the vernacular, and what he calls print capitalism.[10] The printing press liberated the written word from the clergy and the nobility, creating mass communication and a mass audience. In this view, the major element in creating the 'modern' bourgeois state was language: people were French, German, Spanish or English by virtue of their 'mother tongue'. While 'printing press capitalism' did reinforce the vernaculars and enabled literary culture to shift away from the classical language, Latin, the eventual transformation—for example of France—from isolated communities speaking a wealth of tongues to a common French identity took place only between 1870–1914.[11] Railways, better roads, urbanisation, migration, and the need for a common *lingua franca* in the French Army also drove the move towards monolingualism. The French army needed a common language so that soldiers would understand the commands of their officers. A combination of the proto-national identities that Eric Hobsbawm talks about, with the technology of printing and railways, together created modern national identities, but only in a few countries in Western Europe.[12] The key element in this creation of nationalism is its mass character: nationalism has always been an instrument of mass mobilisation, either by the state or by those demanding a nation state.

The physical basis of nationalism is the expansion of

[10] Benedict Anderson, *Imagined Communities: Reflections on the Origin and Spread of Nationalism*, Verso, 2006.
[11] Eugen Weber, *Peasants into Frenchmen: The Modernisation of Rural France, 1870–1914*, Stanford University Press, 1976.
[12] Eric J. Hobsbawm, *Nations and Nationalism Since 1780: Programme, Myth, Reality*, Cambridge University Press, 1992.

Building a Nation with a Scientific Vision 229

communications—mass communication—among a people, and of travel. Railways and the printing press connect people in a shared vision that extends beyond their villages or towns. However it is the mass mobilisation of people, centred on proto-national identities—quite often multiple and even conflicting identities—that creates nations.

Much of the existing literature on the subject, including from the left, focusses primarily on European nationalism.[13] Benedict Anderson examines Latin America and South East Asia as well, but even in his writings, while Creole nationalism in Latin America, Tagalog and Bhasha in Philippines and South East Asia are acknowledged, the national consciousness remains based largely in the vernacular created by print capitalism. The only one who looks at movements creating nations in Africa is Frantz Fanon, but his premature death (in 1961) robbed us of a more complete analysis.[14]

No description of European nationalism is complete without addressing the exclusionary vision of blood and soil nationalism advanced by the ideology of White Supremacism. Instead of including all citizens in the nation, this outlook identifies a majority for inclusion and a minority to be zealously excluded. Nazi

---

[13] Lenin, Rosa Luxemburg, and Stalin have all written extensively on the national question. Perhaps the most quoted of Lenin's texts is *The Right of Nations to Self-Determination* (1914), available as a part of 'Lenin on the National Question', Marxist Internet Archive.
https://www.marxists.org/archive/lenin/works/subject/nation/index.htm
Rosa Luxemburg wrote a series of articles on the same subject:
https://www.marxists.org/archive/luxemburg/1909/national-question/index.htm
Stalin wrote a book, *Marxism and the National Question* (1942):
https://www.marxists.org/reference/archive/stalin/works/1913/03.htm

[14] Frantz Fanon, *The Wretched of the Earth*, Grove Press, New York, 1963. Fanon's preoccupation with the use of force—or violence—in creating the national consciousness during the Algerian War of Independence (1954-62) and his contempt for the nascent African bourgeoisie, who sold their country's wealth to foreign capital after acquiring political power, coloured his vision of the larger issue of how struggles, not just revolutionary violence, create national consciousness.

Germany excluded Jews and the Roma; its modern counterparts, today's White Supremacists, misuse genetics to support their bogus theories of racial superiority.

This form of racial nationalism is not new. It was the basis of proclaiming the White Man's Burden, the thankless task nobly shouldered by white people, who brought civilisation to various benighted others. What befell these others by way of plunder, forced labour and genocide was reduced to a trivial detail. Even European countries like Germany and Italy, who had lost out on acquiring colonies of their own, could participate in white glory by turning their attention to enemies within. Just as the printing press had expanded the sphere of mass communication, the rise of fascist Italy and Nazi Germany coincided with a major expansion of mass communication: the age of the radio and the movie camera. Mussolini's use of radio and Hitler's use of the moving image—Leni Riefenstahl and her Nazi propaganda films, *The Victory of Faith* (1933), *Triumph of the Will* (1935), *Olympia* (1936)—again show that a technology of mass communication does not create fascism, the people using it do. It was the exclusionary movements based on race that created Nazi Germany, not the movie camera of Riefenstahl. For an explanation of the success of such movements, we need to look at economics, history and politics, not just technology. In the same way as the internet today is not a sufficient explanation for the recurrence of majoritarian politics in large parts of the world.

THE BOUNDARIES OF NATIONS ARE CREATED BY MOVEMENTS AND NOT TECHNOLOGY OR LANGUAGES

The first pan-Indian anti-colonial struggle, involving what are now Pakistan, Bangladesh and India, took place against the British in 1857.[15] Colonial historians have dismissed it as the revolt

---

[15] See Prabir Purkayastha in Murli Manohar Prasad Singh and Rekha Awasthi (eds) *1857: Baghawat Ke Daur Ka Itihas*, Granth Shilpi, New Delhi, 2009, pp.

of an unhappy feudal class who had lost power to the corrupt and incompetent English East India Company. All Indians recognise it as a nationalist upsurge, the only discussion being over whether it should be termed the *first* battle for independence.

While the Company's army was a multicultural formation with the need for a *lingua franca*, prior to the revolt of 1857 there was already a dense pan-Indian network of information and communication. Printed matter from modern presses had circulated since the sixteenth century, while the Indian Post Office set up by the Company was opened to the public in 1774. News travelled across the country by more traditional means as well, via itinerant preachers and performers, but largely through the post (*daak*) and newspapers (*akhbarat*). Newspapers were printed both in English and in regional languages. Once control passed from the Company to the British Crown in 1858, a consolidated British Indian Army came into being. Company troops, such as the Bengal Army, had already been used widely in wars fought in India and abroad. The Bengal Army drew its soldiers from a large hinterland, including areas that were to figure prominently in the revolt. Urdu, starting from its Dakhni version in 13th–14th century, was used widely in the military camps. The lingua franca of the Bengal Army was Urdu and Hindustani, at the time they were virtually the same language written in different scripts. The proclamations of the 1857 revolt were in both Hindustani and Urdu.[16] Therefore, the major drivers creating a common identity, at least in large parts of Bihar, UP, Punjab, CP, etc., were already in operation: a common language and the print capitalism emphasised by Benedict Anderson.

Back in Britain, accounts of the revolt focussed on the killing of British women and children by sepoys, but the British themselves engaged in wholesale reprisals, particularly in towns, and had no

---

129–135.

[16] Irfan Habib, 'The Coming of 1857', in *Indian People in the Struggle for Freedom*, Sahmat, 1998.

compunction about putting entire localities to the sword—men, women and children. Mass killings, public hangings, blowing rebel soldiers from the mouths of cannon, mass public humiliations, these were repeated in region after region. By providing Indians this shared memory of brutal oppression, the British handed to the emerging freedom movement a major instrument for shaping its national consciousness. Zafar's pining for '*do gaz zamin*' for his burial in the motherland would symbolise the helplessness of a subject population, and become its rallying cry.[17] In his exile and death, Zafar became much more a symbol of an Indian nation than he was during the uprising.

The question of whether or not India had a national consciousness in 1857 is relevant only if we believe that a national consciousness must predate a movement for its realisation. If we believe, as I do, that the idea of the nation is created through the process of a national movement, then 1857 is the start of this process. That the revolt did not spread to the West and the South is also no argument against it being considered a nationalist uprising. It is the task of the national movement to 'create' both the boundaries of the nation and the 'community' within this boundary that constitutes the future nation. The mass revolt of 1857 is the start of a journey, not its end. As a senior British officer, Thomas Lowe admitted, 'The infanticide Rajput, the bigoted Brahmin, the fanatic Mussalman, and the luxury loving, fat-paunched ambitious Maharattah [sic], they all joined together in the cause; the cow-killer and the cow-worshipper, the pig-hater and the pig-eater'.[18] The mass revolt had indeed united a very large cross section of the people!

[17] *Kitna hai badnaseeb Zafar dafan ke liye*
Do gaz zameen bhi na mili koo-e-yaar mein
How unlucky is Zafar! For Burial,
Even two yards of land were not to be had, in the lane of the beloved

[18] Shamsul Islam, '1857 War of Independence ... when Hindu-Muslim Separatism, Hatred Wasn't an Issue', *Counterview*, 13 May 2020. https://www.counterview.net/2020/05/1857-war-of-independence-when-hindu.html

Building a Nation with a Scientific Vision

The print media and railways enlarged the mass communication space in India, which in turn created multiple larger and overlapping identities: national, linguistic, and religious. Which of these identities would prevail would depend on the mass movements within these boundaries. The boundaries of British India were much larger than what came to be created as its successor states: India, Pakistan, Burma, and Sri Lanka. Again, mass communications—print and railways—may have created the potential for *a nation or nations,* but the eventual boundaries of what emerged were defined by the struggle against the colonial British Indian state. Hence, the anti-colonial struggle against the British creates India. Technology provides the basis, but it is the anti-colonial movement that provides the content of the nation state.

It is not India alone which went through this process. The Arabic language carried the possibility of a pan-Arab state, but the various Arab states that emerged at the end of colonial rule were created by anti-colonial movements specific to each region.[19] The key element in such national struggles is uniting all the people against the colonial oppressor. This is why Indian nationalists focussed on promoting economic nationalism: we needed political independence to develop our economy *for all our people.* The British, they explained, are continuously draining our people and that is why they are rich and India is poor. India's freedom movement was based on civic nationalism—all people within the boundary of India belong here and will become citizens in an independent India.

But there was more than one vision for India even as the struggle for freedom was waged. The creation of a national movement against the British had to contend with a diametrically different form of nationalism.

---

[19] Prabir Purkayastha, paper presented at the conference 'The Spirit of Frantz Fanon', Algeria, 2012.

THE RSS VIEW OF NATIONALISM

For right-wing formations in India, like the RSS, the British were not the enemies; the nationalists, the secularists and the Muslims were. Their view of the nation—held by Hindutva ideologues such as V.D. Savarkar, K.B. Hedgewar, and M.S. Golwalkar—looked to the ideology of exclusionary, ethnic nationalism in Nazi Germany (or Fascist Italy) for inspiration. They contested the inclusive vision at the core of the independence movement and argued that the Indian nation should be based not on its desire to be free from colonial rule, but on 're-discovering an ancient nation'. This is the vision Savarkar formulated, a Hindutva nation based on race, culture, and language: race as Aryan, culture as Hindu and language as Sanskrit. Savarkar himself clarified that the Hindutva on which he based the nation was different from the Hindu religion.[20] The RSS leaders took over this concept of the nation from Savarkar. Such a nation, said Golwalkar, was based 'on essential value of the five unities, Country [Geography], Race, Religion, Culture and Language towards making a complete Nation concept'.[21] For them—as with their descendants such as Modi—foreign rule began with Muhammed Ghori's victory over Prithviraj Chauhan.[22]

What is the link between Savarkar's exclusionary nationalism and its forms and manifestations elsewhere? All these different forms of exclusionary nationalism define the state on the basis of race, religion, or ethnicity. Using a particularly toxic form of nationalism as their ideology, these exclusionary nationalisms

---

[20] Vinayak Chaturvedi, 'Reading Savarkar: Was the Hindutva icon actually Hinduphobic?' *Scroll.in*, 06 September 2021. https://scroll.in/article/1004641/reading-savarkar-was-the-hindutva-icon-actually-hinduphobic

[21] M.S. Golwalkar, *We, or Our Nationhood Defined*. 1939, p. 116.

[22] In his maiden speech in the Lok Sabha, Modi referred to 1,200 years of servitude, in Hasan Suroor 'Chaining 1,200 years', *Outlook*, 7 July 2014. https://www.outlookindia.com/magazine/story/chaining-1200-years/291200

Building a Nation with a Scientific Vision

are put into practice by supremacists. So white 'nationalism' is racism, just as Christian, Hindu, or Islamic 'nationalism' involves supremacists of varying hues using religion as their cover. Examples of these supremacy-driven nationalisms abound in Europe, especially in the long historical process that led to the rise of nation states.

This process was accompanied by external wars to establish national borders and acquire colonies.[23] The colonisers' vision of their home country, the 'true nation' based on exceptionalism and racial superiority, came to be reflected in the blood and soil vision of Savarkar-Golwalkar's Aryan Hindu nation. How did colonised subjects such as Savarkar and Golwalkar decide to use the very concepts that had led to their subjugation? Was it that the Brahmanical world-view, with a caste supremacist vision of India, made it easier to assimilate a racist outlook?[24]

Savarkar's concept of an Indian nation was not an original one, but borrowed from late nineteenth and early twentieth century German nationalism, with its basis in blood (race) and soil, which Nazism had appropriated.[25] Blood and soil are recurring themes in Savarkar's writing on the nation.

THE INCLUSIVE NATIONALIST VISION

In the period of struggles against colonial regimes, nationalist forces all over the world—in India as much as elsewhere—had differing views of the nation. It is not surprising that in most national movements, civic nationalism—with its core of economic

---

[23] Alsace-Lorraine, now called Alsace-Moselle, saw three wars between France and Germany, the Franco-German War 1870-71, and two world wars.

[24] Raosaheb Kasbe's book *Jot* in Marathi, now translated into English as *Decoding the RSS*, exposes the deeply casteist views of the RSS founders. Not surprisingly, the RSS can't stand this book. When it was first published in Marathi, RSS cadres made a public bonfire of it in Pune. Raosaheb Kasbe, *Decoding the RSS: Its Tradition and Politics*, LeftWord Books, 2019.

[25] Johann Kaspar Bluntschli, *The Theory of the State*, in German 1875, in English translation, 1895.

Figure 11.1: A History of World GDP

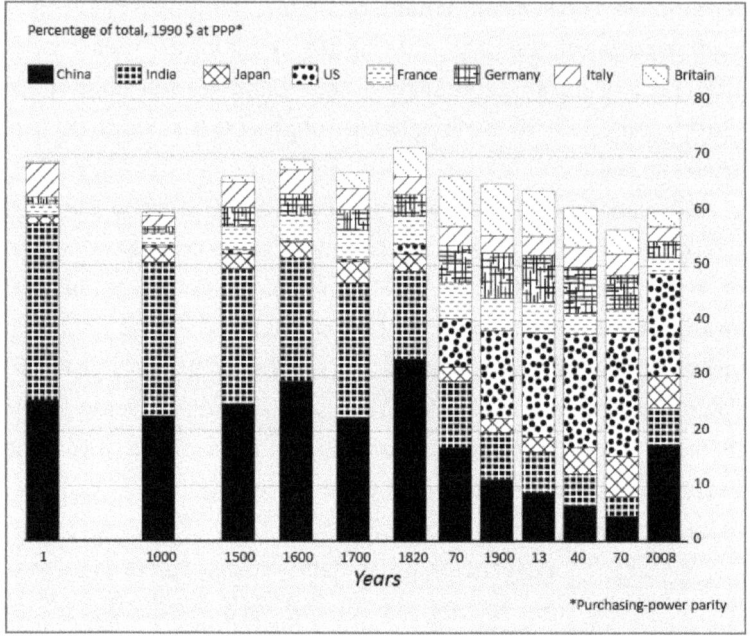

Source: Angus Maddison, *Contours of the World Economy, 1-2030 AD*, Oxford University Press, December 2007; Kenneth N. Cukier 'More Than 2,000 Years in a Single Graphic', *The Economist*, 20 June 2012. https://www.economist.com/graphic-detail/2012/06/20/more-2000-years-in-a-single-graphic.

nationalism—became the dominant force, since the economic exploitation of people and countries was at the heart of colonialism.

The colonial conquerors had looted, enslaved, and massacred the people of the Americas, Africa, and Asia on a grand scale, and finally built a system that continually created wealth in the Western metropolis while impoverishing the colonies. This is why, as we can see in Figure 11.1 (from Angus Maddison's classic work), India and China, which till the eighteenth century had produced about 50 per cent of the world's GDP, came down to less than 10 per cent within the next 200 years.[26]

[26] Angus Maddison, *Contours of the World Economy, 1-2030 AD*, Oxford University Press, December 2007.

Building a Nation with a Scientific Vision

However, this was not just due to the draining away of tribute and plunder from the colonies or semi-colonies. Imperialism created a system that led to the continuous development of productive forces by harnessing science and technology, while bleeding the colonies. Alongside the increasing production of goods unleashed by the industrial revolution, there was colonial extraction—transferring raw materials from the colonies.[27] It also destroyed the manufacturing industries in the colonies, converting them into captive markets for the sale of goods from factories in the metropolis. After the industrial revolution, Britain wanted only raw materials from its colonies, *not any industrial products*: the slogan was 'not even a nail from the colonies'. Technological and financial initiative, as much as governance and regulation, was established as the unique province of the coloniser, while the colony became a helpless dependent.

AN INDEPENDENT NATIONAL ECONOMY AS THE BASIS OF THE NATION

Early nationalists such as Dadabhai Naoroji (1825–1917) saw how colonial rule bled India while enriching Britain. Naoroji's 'drain of wealth' theory was among the earliest attempts to analyse colonialism economically, establishing the link between colonial rule and poverty in India.[28] The eventual challenge before Indian nationalists was how to get the postcolonial state to act on a vision that embodied two complementary tasks: an India for all of its people, and a state that would develop all its resources—including its human resources.

---

[27] According to research by economist Utsa Patnaik, a total of $44 trillion were siphoned off from India by colonial Britain in Shubhra Chakrabarti and Utsa Patnaik (eds.) *Agrarian and Other Histories: Essays for Binay Bhushan Chaudhuri*, Tulika Books, 2018.
https://cup.columbia.edu/book/agrarian-and-other-histories/9789382381952

[28] Dadabhai Naoroji, *Poverty and Un-British Rule in India*, 1901.

It is useful to note the contrast between these two visions of the nation, the inclusive nation of the independence movement, and the divisive blood-and-soil European version that fathered Golwalkar and Jinnah's vision. These two versions are not specific to India, but are part of a larger struggle in the national liberation movements of the colonies.

In India, leaders such as Nehru, Ranade, and Gokhale located the nation on the terrain of an independent national economy.[29] The starting point of the movement against imperialism was the struggle against the colonial state, which united the people and shaped the anti-colonial national consciousness. It formed the central impulse of the independence movement against the British. These two elements—secularism as the basis of uniting the people, and economic development—distinguish the vision and action of all the leaders of the independence movement, whether Patel, Bose, Ambedkar, Nehru or others.

Indian nationalists knew that the British were different from earlier conquerors. In the early sixteenth century, Babur had understood that once he was ruling India there was no way he could go back to his beloved Fergana Valley. India went on to absorb the Mughals, as it did all its earlier conquerors. They became as much a part of this land as any others. By contrast, in the Hindutvavadi view, the religion of the ruler decides whether India is considered to be under foreign rule, even when such rulers were yet another set of migrants, such as the Aryan migrants who Savarkar admits came from outside and settled on the banks of the Indus. Savarkar separates these migrants to South Asia from all other migrants, by claiming that the land imbued the Aryan 'race' with its holiness, as their religion developed here—along with their 'holy' language, Sanskrit.

Savarkar and Golwalkar lacked the understanding that a coloniser's view of what constitutes a nation can only form the

---

[29] Manu Goswami, *Producing India: From Colonial Economy to National Space*, University of Chicago Press, 2004.

Building a Nation with a Scientific Vision 239

basis of a colonial or colonised nation. The European colonisers had wanted foreign territories and had colonised people in order to plunder their wealth and resources. Through slavery—which formed a big part of the colonial plunder—they bled the subject people of even their humanity. Using such a version of nationalism as inspiration could hardly provide colonised people with the emancipatory vision that a free nation, and a free people, require. It would certainly not alter their conditions of life in any emancipatory sense.

BUILDING THE VISION OF THE NATION THROUGH THE CONSTITUTION: THEN AND NOW

India became a Republic on 26 January 1950, when its new Constitution came into effect after being steered through the Constituent Assembly by Dr B.R. Ambedkar. The Constitution is what guarantees to all sections of our people—irrespective of race, religion, gender, or caste—full rights to the nation, including the right to a decent standard of living. This vision of democracy and a secular republic has come under threat today from those sections that never joined the freedom movement and had a much narrower, sterile vision of the Indian nation.

The Indian Constitution was not simply a product of debates in the Constituent Assembly. It reflects the values of the freedom movement which had united people of different religions, languages and ethnicities, even ideologies, in a common struggle against the colonial oppressor. The freedom movement wanted an end to colonial rule and, with it, to the *colonial exploitation of India*. An independent India had to deliver to all its people freedom from both foreign rule and from the grinding poverty this rule had imposed on them. This inclusive vision gave rise to the world's first large-scale affirmative action: reservation for certain sections—dalits and adivasis—who had been excluded from caste Hindu society and had suffered generations of exploitation and violence.

Of course, this unity of goals—ending colonial exploitation and establishing a secular state—had other fault lines. In his speech to the Constituent Assembly on 25 November 1949, Ambedkar said:

> On the 26th of January 1950, we are going to enter into a life of contradictions. In politics we will have equality and in social and economic life we will have inequality . . . How long shall we continue to live this life of contradictions? How long shall we continue to deny equality in our social and economic life?[30]

Sections of the left opposed the Constitution as the right to property had been included as a fundamental right. The Supreme Court needed decades to establish the doctrine of the 'basic structure' of the Constitution, and to determine that rights which do not belong to the basic structure—such as property rights—could indeed be amended by the Parliament.[31] Despite such points of difference between Ambedkar and the left, both worked within the larger vision of an inclusive state that belonged to, and had, therefore, to deliver development to all its people. This vision was not shared by either the Muslim League or the Hindu Mahasabha-RSS, who wanted a state based on religious identity and for capital—both foreign and local—to develop the economy, with the state playing only a minimal role. Jinnah wanted Pakistan in the belief that an independent India that had Hindus as the majority would become a majoritarian state. Independent India explicitly rejected this vision of a majoritarian state—a Hindu Pakistan, as Nehru called it—choosing instead a secular and inclusive Constitution.

The Constitution and the Republic that we celebrate were

---

[30] B.R. Ambedkar, 'The Grammar of Anarchy', Speech to Constituent Assembly, 25 November 1949.
https://speakola.com/political/babasaheb-ambedkar-grammar-of-anarchy-1949

[31] Kesavananda Bharati & Ors. v. State of Kerala & Anr., 1973.
https://indiankanoon.org/doc/257876/

never accepted by the RSS and its various front organisations. While they might have decided today that it is better to try and subvert the Constitution slowly, they never hid their opposition to its fundamentals. The RSS and the Jana Sangh had derided Ambedkar as 'a Lilliput' (sic), while extolling Manu and the *Manusmriti*. *The Organiser*, mouthpiece of the RSS, stated in its issue of 30 November 1949:

> The worst about the new Constitution of Bharat is that there is nothing Bhartiya about it. The drafters of the Constitution have incorporated in it elements of British, American, Canadian, Swiss and sundry other constitutions.... But in our Constitution, there is no mention of the unique constitutional development in ancient Bharat. Manu's Laws were written long before Lycurgus of Sparta or Solon of Persia. To this day his laws as enunciated in the *Manusmriti* excite the admiration of the world and elicit spontaneous obedience and conformity. But to our constitutional pundits that means nothing.

It is not surprising that the other development of the Indian Constitution, of affirmative action or reservation in education and employment, was also opposed by the RSS. Even today, RSS leaders speak out against reservation: how it promotes separatism and why it should be wound up.[32]

STATE PLANNING, SCIENCE, AND THE PUBLIC SECTOR

The national movement was aware that freedom would not be truly achieved till India was free from the absolute poverty, abysmal life expectancy, and illiteracy that British colonial rule had left behind. The country had to act on a vision that embodied two

---

[32] Amid the continuing Patel quota stir in Gujarat, 2015, RSS chief Mohan Bhagwat asked for a review of the reservation policy. He repeated this demand in August 2019.

complementary tasks: a state that would work for all its people and develop all its resources, including its human resources. For this to happen, different sections of the independence movement—from the leaders on the left to Nehru, Ambedkar, and Bose—were united in the vision that we needed science and technology to develop the country's productive forces.

The Planning Commission, and its precursor, the Congress Planning Committee, embodied this vision of India's development. After independence, along with planning the economy and building the country's infrastructure, scientific institutions such as the Council of Scientific and Industrial Research and the Indian Council of Agricultural Research were strengthened to help India develop indigenous technology. Higher education was expanded, including science and technology institutions; and new institutions such as the Indian Institutes of Technology were set up.

The leaders of the independence movement knew that science cannot be borrowed or bought. It needs to be developed indigenously. Without scientific and technological knowledge, neither industry nor agriculture can develop. They also understood that the development of a newly independent nation required the adoption of a scientific outlook on both nature and society. This would help people shed the shackles of superstition. Scientific temper, or a scientific outlook towards nature and society, is how we develop productive knowledge for a new future.

Bose, as the Congress President, set up the Planning Committee in 1938,[33] which he asked Nehru to head. Both drew inspiration from the Soviet experiments with planned development after the October Revolution. After independence, the Planning Commission carried forward the vision of the Planning Committee to overcome the double burden of poverty and inequality left behind by the British. The leaders of the national movement viewed planning and building a public sector as a necessity, not

---

[33] *Subhas Chandra Bose, Pioneer of Indian Planning*, Planning Commission, 1997.

Building a Nation with a Scientific Vision   243

just for the industrial and agricultural regeneration of India, but also to re-distribute the benefits of development to all sections of its people. They wanted to develop the nation's productive forces based on scientific knowledge, and the biggest resource for a nation developing its people's scientific capability was education.

Developing scientific and technological capabilities became a priority for the Indian state. It built the Central Scientific and Industrial Research laboratories, the five Indian Institutes of Technology, the Indian Statistical Institute and a host of other scientific institutions. Beginning with the Bombay Plan formulated by Indian capital in 1944–45, it was common agreement that development needed infrastructure, and only the state had the capacity to develop infrastructure on the scale that India required for rapid development.[34]

Many identify Nehru as the creator of the hydro-electric projects India undertook after independence. While Nehru fully supported multi-purpose irrigation and hydro-electric projects, people forget that the blueprint of multi-purpose projects—combining irrigation and electricity generation—as well as a unified national grid, was developed by Ambedkar. As the Law Minister, Ambedkar also drafted and introduced the Electricity Act, 1948. He believed that developing electricity generation and a national grid was the basis of industrialisation, and saw 'industrialisation as the surest means to rescue the people from the eternal cycle of poverty in which they are caught'.[35] This vision of electricity as a necessity (not a market commodity to make profits) formed the basis of Ambedkar's Electricity Act. After independence, the

---

[34] Amal Sanyal, 'The Bombay Plan: The Industrialists Behind India's First National Economic Plan', *Quartz India,* 15 November 2018. https://qz.com/india/1464869/the-story-of-jrd-tata-gd-birlas-bombay-plan-for-india

[35] Ambedkar's presidential address to the Policy Committee on Electric Power in Sukhdeo Thorat (ed.) *Ambedkar's Contribution to Water Resources Development,* p. 50, Central Water Commission. 1993. http://cwc.gov.in/sites/default/files/ambedkars-book_1.pdf

Indian state did not just build the public sector but also invested in people. When the Damodar Valley Corporation was created, more than fifty engineers were sent to the Tennessee Valley Authority. They formed the core of the Indian power sector. Again, a group was sent to US Steel to learn about the industry; they became the core leadership of the Steel Authority of India. The history of the IT sector shows the role the Indian Institutes of Technology, the Indian Statistical Institute, the Electronics Corporation of India Ltd., and other public sector bodies have played in its explosive growth.

India used the state to build its infrastructure. This vision was embodied in successive Five-Year Plans. In the game of posing Ambedkar versus Nehru or Subhas Bose versus Nehru, the BJP would like us to forget what these figures stood for. Yes, these leaders had their differences about tactics, about international issues, the timing of struggles, and so on. But what united them was a larger vision of a secular, democratic India that needed science and technology for development, a state which had to take along all sections of the people.

The vision was destroyed jointly by the BJP and Congress, both supporting the neoliberal paradigm of market-knows-best, and advocating the withdrawal of the state. This was the underpinning of the new Electricity Act of 2003. That the Congress should have laid the groundwork to dismantle the electricity sector though their policies of privatisation and fragmenting the grid speaks volumes for the road they have travelled from the independence movement and its vision of India. As for the BJP, they never supported Nehru's or Ambedkar's view of planning and the economy. So their view of the electricity sector is no surprise. Moreover, the rise of the BJP has also seen Hindutva's vision of ethno-nationalism take a contemporary form in times of neoliberalism.

Building a Nation with a Scientific Vision
THE RETURN OF ETHNO-NATIONALISM IN NEOLIBERAL TIMES

The rise of the ideology of neoliberalism has been co-terminous with the weakening of the nation state. All the political parties that called themselves liberal, socialist, or communist, while differing widely on their goals, had one idea in common: they all saw the state as the instrument of redistribution of wealth. The liberals believed in education, health, and other social objectives requiring state intervention, and therefore in taxing the rich. It was the liberal parties in United States, for example, that broke up monopolies. Social democracy in UK, France, and West Germany, introduced public health and subsidised public education. Communist parties in the Soviet Union, China, Vietnam, not only used the state for re-distributive functions but also for the development of the economy. As did newly independent countries such as India, Algeria, Ghana, and Egypt.

The weakening of the state in its redistributive role took place with the fall of Soviet Union and the socialist countries in Eastern Europe. Capital, threatened by the rise of the socialist countries, no longer felt compelled to grant concessions to either its working class or the developing countries. With the socialist bloc disintegrating as the other pole of international politics, Big Capital centred in the US-Western Europe was free to re-colonise the world. Not in the old way, with political control—for they still needed the developing countries to be led by their ruling classes—but in tandem with the neoliberal ideology of free flow of capital and goods. This is the ideology that the World Bank and the World Trade Organisation—established in 1994—bring to the world. Nations should be politically free but economically subordinated to global capital.

The advance of the neoliberal agenda with its demand for the free flow of capital and commodities, is generally seen as unrelated to the fault lines of ethnic and religious violence that are opening up in various countries. If a relationship between these two phenomena is admitted, it is largely in terms of an assault by the forces of backwardness on a 'civilised West'. Even the victims of this assault do not challenge what constitutes Western civilisation but only invert its categories: claiming a past in which they, not the West, were superior. To others, the nation state is passé, an anachronism to be overcome in an increasingly globalised world.

Concepts framed in economic language are not quite as they seem: often enough, terms are used to mean their opposite. Globalisation, competition and free trade are some of these terms. People may think they are all positive ideas: globalisation should mean the free flow of people and ideas, allowing advances taking place anywhere to be shared by others. Competition should translate as more choices for the consumer, and free trade should make it easy for producers to get a better deal. Unfortunately, the reality is quite different.[36] Globalisation has meant a free flow of commodities and capital, increasingly speculative capital, but not of ideas or knowledge, which come up against a much harsher intellectual property rights (IPR) regime. Competition, in practice, has meant mergers and acquisitions on a global scale and the creation of mega monopolies.

The structure of this globalisation was created in the era of WTO and TRIPS (signed in 1994). But this was not the only instrument that global capital used. It was buttressed by a series of free trade and investment agreements that have effectively made the laws of countries subordinate to the country from where the investment is channelled.[37] The other element of this picture is

---

[36] Benjamin Selwyn and Dara Leyden, 'World Development Under Monopoly Capitalism', *Monthly Review*, Vol. 73, 6 November 2021.
https://monthlyreview.org/2021/11/01/world-development-under-monopoly-capitalism/

[37] Kavaljit Singh, 'Treaties That Gave Away the Store', *The Hindu*, 27 April 2012.
https://www.thehindu.com/opinion/lead/treaties-that-gave-away-the-store/article3357429.ece
Kavaljit Singh and Burghard Ilge, *Rethinking Bilateral Investment Treaties: Critical Issues and Policy Choices*, Both Ends-Madhyam-Somo, 2016.
https://www.somo.nl/wp-content/uploads/2016/03/Rethinking-bilateral-

that the current form of globalisation is meant for finance capital and commodities but not for people, who are excluded from the globalisation scheme. They come up against an ever-harsher immigration regime. While a global corporate elite—rich, mobile, and integrated—is emerging, the overwhelming majority in most countries is becoming marginal to the global economy. They are increasingly pauperised as their nations implode or fracture.

Some supporters of globalisation argue that it is better to submit to a US imperium with its 'Western values', as this would automatically secularise countries such as India. With increased globalisation, runs the argument, our world will become infused with common global values, presumably better ones than the national and local variety currently in vogue.[38] Unfortunately for this view, globalisation has been accompanied by far more violent religious and ethnic conflicts: the cohesion between a 'civilising' global order and realities on the ground is breaking down.

THE WEST AS AN IDEOLOGY: SANITISING THE COLONIAL HISTORY OF GENOCIDE, SLAVERY AND PLUNDER

The global neoliberal agenda aims to dismantle the nation state, except for its police functions. As early as 2000, Wayne Ellwood had noted: 'True believers in the neoliberal agenda would prefer a pliant nation state, one which supports them when it is necessary and stands aside the rest of the time.'[39] This is why nation states have become increasingly fragile as they lose legitimacy in the eyes of their people.

If nation states are to be dismantled except for their police

---

investment-treaties.pdf

[38] Today, this is the so-called 'rule-based international order', where the G7 countries or the 'West', a club of ex-colonial and settler colonial states, get to make the rules.

[39] Wayne Ellwood, 'Redesigning the Global Economy', *New Internationalist*, Issue 320. January–February, 2000.
https://newint.org/issues/2000/01/01?page=2

function, they cannot have legitimacy without redefining themselves. A state that only oppresses its people to further global corporate rule must promote some other identity; it is here that the ethnic and religious agenda come into play. By redefining the nation in religious or ethnic terms, political leaders turn the focus away from the process of re-colonisation. For the new imperial order, it is tricky to justify global corporate rule over the world, their own people included. Therefore, they need the legitimising myth of a rational and civilised 'West' being confronted by authoritarian forces of darkness rising out of the 'non-West'.

Since 11 September 2001, a recurring motif in the media is that of a clash of civilisations. The West and Islam are pictured in perpetual conflict—going back to the Crusades—apart from a brief interregnum when the West had clashed with the Evil Empire of Communism. We have heard expert after expert, and political leaders—dim-witted or otherwise—echoing that the attack on the World Trade Centre was an attack on Western values and civilisation. The underlying message was clear: the killing of thousands of innocent people is utterly non-Western and could be the handiwork only of those who find themselves incompatible with the West. The colonial massacre of Jallianwala Bagh, Churchill's Bengal Famine, the brutal suppression of the Kenyan people, and the colonial battles in Africa are conveniently forgotten in this retelling of history. Also airbrushed out of history is the West's record of slavery, genocide, and loot on which its 'civilisation' is based. Along with the many coups against democracy it has engineered in countries from Guatemala and Iran to Chile. Or the dictators and theocrats who remain in power solely by the grace of the West, rather than the will of their people.

It is not only the 'injured' West that endlessly replays the theme of a clash of civilisations. The Indian (Hindutva) variant has projected an alliance of civilisations—Hindu-Christian and Zionist—against the new Evil Empire of resurgent 'jihadi' Islam,

Building a Nation with a Scientific Vision 249

and portrayed itself as a victim of past Islamic dominance, seeking redress.

If this alleged global conflict of religions is cloaked in the West as the Huntington thesis—'clash of civilisations'—a Talibanised Islam echoes this view. And back home we have the Hindutva lobby, seeking to project a homogenised monolithic Hinduism against an equally monolithic Islam, claiming 1,200 years of living in servitude, not the 200 years of British rule.[40]

The subtext of such worldviews is the idea that our current society is built on a foundation of hate and conflict: 'we' and 'they' being interchangeable as we move across the world. Periodic sectarian clashes—Israel and Palestine, Bosnia, Afghanistan and, closer to home, Gujarat—are a reaffirmation of this framework of a clash of religions.

Myth-making and falsifying history is a part of this project towards the future. History is crafted selectively and quite often fabricated. For example, in this version of the West, Greece—the classical civilisation—is the progenitor of all Western values. And this Greece arose from a primordial Aryan Greece with hardly any influence of the proximate older civilisations of African Egypt, Semitic Phoenicia, or Anatolia. A deeply racist view of history underpins the popular perception, a manufactured past in which the 'West' and 'Western values' have borrowed nothing from elsewhere. All other groups reside in spaces informed by values that are largely irrational and primitive, manifestly inferior. The bloody wars among nations in Europe are all forgotten. They somehow become extrinsic to the innate 'West' that germinated in Greece, lay dormant in the Dark Ages, and has now sprouted green shoots all over the physical West—Europe and the Americas—and the physical non-West of Australia and New Zealand. That

---

[40] Debobrat Ghose, '1,200 years of Servitude: PM Modi Offers Food for Thought', *Firstpost*, 13 June 2014.
https://www.firstpost.com/politics/1200-years-of-servitude-pm-modi-offers-food-for-thought-1567805.html

indigenous people in the Americas were 'ethnically cleansed' (just like the Moors of Spain, after being in the Iberian Peninsula for more than 800 years), does not find place in this narrative.

If mythmaking did not also involve building a particular future, we could perhaps ignore it as common prejudice that would disappear with a better appreciation of the past. However, in the way that colonialism sought legitimacy by defining itself in terms of its civilising role, current imperialism couches its mission in terms of spreading reason. Its account of Western History as Reason and Civilisation is a myth, and exposing this truth is crucial in the fight against the neo-imperial order.

THE BJP VISION: A LETHAL MIX OF EXCLUSIONARY POLITICS AND NEOLIBERALISM

Development today is not just the development of factories and machines but also of the knowledge that is embedded in the machines. To develop technology, independent India set for itself the goal of self-reliance, or 'Made in India'. The goal of self-reliance in technology was backed by a set of policies that insisted, in any foreign partnership, on the transfer of all technology to the Indian entity. In this policy of self-reliance, transferring knowledge was as important as imported plants and machinery.

In contrast, we are now witnessing a continuous assault on institutions of education and research, and on reason and science; myths and madness masquerade as science and history—flying chariots and interplanetary travel, genetics in the Mahabharata, talk of evolution as false or, if true, superseded by the 'much superior' theory of *dashavatar*.[41] The objective is a 'nationalist' India based on

---

[41] Andhra University Vice-Chancellor G. Nageshwar Rao speaking to the 106th Indian Science Congress, January 2019. PTI 'Theory of Evolution from Dashavatar Superior to Darwin's: Andhra University VC', *Firstpost*, 5 January 2019.
https://www.firstpost.com/tech/science/theory-of-evolution-from-dashavatar-superior-to-darwins-andhra-university-vc-5842761.html

Building a Nation with a Scientific Vision 251

religious identity. Hence the need to demolish reason and history, so as to generate a majoritarian India in which minorities would have very few rights; an India where reason has to be surrendered to myths old and new; where wealth and caste means merit.

Aligning with the imperialist, capitalist powers and not supporting national liberation movements was also the *post-independence* foreign policy view of the RSS: non alignment and planning were both viewed as two sides of the same evil, socialist coin. Instead, they argued for a 'holy' alliance of Christians—read ex-colonial powers and the US—Jews (read Zionist Israel) and Hindus on one side, against the unholy communists and Muslims.

The RSS was bitterly opposed to planned development and the public sector. They wanted India to be completely left to market forces, along with unfettered entry for global capital. The only role for the state was to help Indian capital negotiate with foreign capital; in other words the crony capitalism we see in action today. This is why Modi replaced the Planning Commission with a toothless think tank called the NITI Aayog. It is why he is dismantling the public sector, selling it to friendly capitalists, and inviting foreign capital under the slogan of Make in India. It is a journey of betrayal—from self-reliance to Reliance!

The difference between the idea of genuine self-reliance, and the current vacuous slogan of Make in India is this: one involves insistence on the transfer of knowledge and developing that knowledge further; the Modi version is an invitation to global capital to exploit India's cheap labour, along with various tax breaks and subsidies, including virtually free land.

In Hindutva's exclusionary view of nationalism, it is the land that is the nation, the land that is pure: Savarkar's *punya-bhumi* and *pitru-bhumi*. That is why Modi—quoting Deendayal Upadhyaya—said in September 2016 that Muslims have to be purified (*parishkar*) to be fully Indian.[42]

---

[42] At Deendayal Upadhyaya's birth centenary celebrations, Narendra Modi declared, 'Fifty years ago, Pandit Upadhyaya said, "Do not reward/appease

The striking feature of the Hindutva variety of nationalism is not just what it claims as its basis but equally what it does not. Nowhere does it talk about the economic basis of nationalism: the right of a people to control their economy, market and resources. It is not an accident that the BJP government is quite happy to surrender India's economy to the US, IMF and World Bank, while claiming to be nationalist. 'Hindu' nationalism has no space for developing the national economy, except in handing it over to Indian or foreign capital. Presumably, foreign capital gets fully sanctified as Indian once it reaches the *punya-bhumi* of a Hindu India.

Today, as we give up the hard-won economic space secured in our postcolonial period, the very basis of our civic nationalism collapses—along with the broader vision of development championed by the anti-colonial struggle. Under the assault of neoliberal globalisation, once leadership in country after country began to cede the economic space, they were left with two choices: they could either give up the concept of the nation itself, or define the nation purely in terms of a cultural, linguistic or ethnic identity. As giving up the nation would mean accepting virtual re-colonisation, they settled on the alternative, falling back on a narrow view of nationalism with its attendant ethnic or religious fault lines.

The RSS cannot tolerate a pluralist basis of the nation. The secular definition of the nation, by which the state is neutral between religions and does not allow religion a place in governance, goes against their view of a monolithic nation built on a common identity. Given that such an identity did not exist, the need arose to create one by using religion.[43] At the core of the dispute is the

(puraskrit) Muslims, do not shun (tiraskrit) them, but purify (parishkar) them.'" Christophe Jaffrelot, 'Hindutva's "Purification" Drive', *The Indian Express*, 13 October 2016.
http://indianexpress.com/article/opinion/columns/hindutvas-purification-drive-muslims-india-islam-caste-system-3079478/

[43] Naunidhi Kaur, 'Voices from the Far Right', *Frontline*, 4 August 2001. https://frontline.thehindu.com/other/article30159752.ece

definition of the nation, and in this dispute the fundamentalists and the communal fascists are on the same side.

There is a corollary to this nation constructed on the narrow identity lines imposed by Hindutva. This is the identification with one central idea of neoliberalism: the market, meaning Big Capital, takes charge of the economy, from investment decisions to the price of goods. The state is to withdraw from investing even in critical sectors of the economy, handing them over to Big Capital. While the state can directly distribute some largesse—food, cooking gas, vaccines—especially before the elections, it should let the market decide the price and where investments should take place. In effect, it has meant handing over the entire economy to Big Capital.

The BJP government has wound up the Planning Commission. It has steadily handed over higher education to private and even foreign universities, while involving people without an understanding of science or technology in running advanced institutions. Having the 'right' ideology is, for the BJP-RSS, much more important than the development of knowledge. It might appear that the BJP's contempt for knowledge is dangerous only to the social sciences. The attacks on the Jawaharlal Nehru University (JNU) may be the most visible instance of the BJP's destructive approach; equally, it may seem that JNU is a target because the university has groomed a generation of scholars in the social sciences who are perceived as a threat to the BJP's politics. But the BJP government and its plants in the universities have not limited their attacks to the social sciences or to JNU. Their attack is on knowledge itself. In institution after institution, people with no vision, and very little learning, have been placed in powerful positions. Knowledge has been reduced to a secondary place; what matters is that universities indoctrinate their students with the BJP-RSS ideology.

Giving free rein to fantasy and bigotry impairs the ability to register facts, let alone to work with them. The Modi government does not recognise that people and knowledge are key in technology development today. Of the top five companies of the world—measured by market capitalisation—four are digital monopolies.⁴⁴ Take Apple Inc., which is the biggest company in the world but does not own a single factory. How does it do this? It owns the designs, the software and brand of Apple. Apple gets about $300 for each iPhone it sells, while Foxconn, the company that manufactures the phone gets only about $8.⁴⁵ This is the nature of the knowledge economy.

Not surprisingly, in spite of Modi's hype of Make in India, India's year-on-year GDP growth has been slowing down significantly. Even after the end of the second wave of COVID-19 in 2021, India's GDP was well below the 2019 figure, making a mockery out of soon becoming a five-trillion-dollar economy.

The current attack against minorities and certain castes and communities is not an aberration. It is fundamental to how the RSS, the BJP, and their front organisations think. These attacks are on the fundamental values enshrined in our Constitution, including economic democracy.⁴⁶ The attacks are taking place when India has again become as unequal as it was under the British, or as Thomas Piketty has called it: from British Raj to Billionaire Raj.⁴⁷ Just nine families today own more wealth than half of all Indians.⁴⁸

---

⁴⁴ Dogs of the Dow, '50 Largest Companies by Market Cap Today (TOP 50 LIST)', 23 September 2022.
http://dogsofthedow.com/largest-companies-by-market-cap.htm

⁴⁵ Sarah Mishkin and Maija Palmer, 'Foxconn Survives on Thin Slices of Apple', *Financial Times*, 25 September 2012.
https://www.ft.com/content/170a225c-0356-11e2-a284-00144feabdc0#axzz27UlBuFXg

⁴⁶ Aijaz Ahmad in his essay calls the Modi government in India and Erdogan's in Turkey as 'post-democratic'. 'Extreme Capitalism and "The National Question"', Vol. 55: *Socialist Register*, 2019.
https://socialistregister.com/index.php/srv/article/view/30924

⁴⁷ Lucas Chancel, Thomas Piketty, 'Indian Income Inequality, 1922–2015: From British Raj to Billionaire Raj?', July 2017.
https://wid.world/document/chancelpiketty2017widworld/

⁴⁸ Oxfam Inequality Report, 'Public Good or Private Wealth, the India Story',

It is not where you produce, but what knowledge you have that determines winners and losers in today's global economy. Developing its people is the key to the future development of a country. This is why any nationalism that defines itself through a land, and not its people, belongs to the past. A scientific vision of the past and of the future is key to this fight. Giving up knowledge in the belief that the ex-colonial powers will readily hand it to us is a project for the re-colonisation of India. This is why we have to fight. Our battle is for the sovereign socialist secular democratic republic that we envisioned during the independence movement.

---

21 January 2019.
https://www.oxfamindia.org/sites/default/files/Davos-India_Supplement.pdf

Acknowledgements

This book has benefitted from the work of several individuals, groups and movements over the years. But it would not have been a book at all without the coaxing and bullying of Githa Hariharan; the intelligent and meticulous editing of Salim Yusufji; Winnie Chauhan's sharp eye; and the patient assistance of Abhilasha Chattopadhyay. Venkatesh Athreya and Bappaditya Sinha read the manuscript and offered helpful suggestions. Sudhanva Deshpande offered unstinting support to the idea and the making of the book. I thank them all.

The source of D.D. Kosambi's quotation is from *Exasperating Essays*, People's Publishing House, New Delhi, 1957. https://www.marxists.org/archive/kosambi/exasperating-essays/x01/1952.htm.

Section I: An earlier version of Chapter 2, 'The Knowledge of Science and Technology as Commons', was published in Vocabulary of Commons, Foundation for Ecological Security (FES) 2010, pp. 445–462. https://www.slideshare.net/OpenSpace/vocabulary-of-commons?from_action=save.

Chapter 3, 'The COVID Pandemic Experience: Who Won, Who Lost?' draws partly on my earlier essays, 'COVID-19 Pandemic and the Pathologies of Late Capitalism', *Marxist*, Vol. XXXVI, 2, June 2020. https://cpim.org/sites/default/files/marxist/marxist_2020_2_apr-jun_01_edit.pdf; 'A Caring World Needs to Share Knowledge to End the COVID-19 Pandemic', *Newsclick*, 15 May 2021. https://www.newsclick.in/world-needs-to-share-COVID-19-Vaccine; and 'The West is Practicing Vaccine Apartheid at a Global Level', *Newsclick*, 10 April 2021. https://www.newsclick.in/the-west-practicing-vaccine-apartheid-global-level.

Acknowledgements

Section II: Chapter 4, 'Understanding the Philosophy of Technology' is a lightly edited version of 'Towards an Understanding of the Philosophy of Technology', published in *Water World*, Vol 1, No 1, July-September 1978, pp. 20–23.

Similarly, Chapter 5, 'Restoring Conceptual Independence to Technology', updates an article in the *Economic and Political Weekly*, Vol. 37, 1, Perspectives, pp. 33–37, 5 January 2002. https://www.epw.in/journal/2002/01/perspectives/restoring-conceptual-independence-technology.html.

Chapter 6, 'The Dynamics of Technology and Self-Reliance' draws on 'Technology, Self-Reliance and Public Domain Science' in *Social Scientist*, Vol. 31, No. 11/12, November-December 2003, pp. 86-99. https://doi.org/10.2307/3517951.

Section III: An earlier version of Chapter 7, 'Science in the Light of Social History', was published as 'Science, History and Society' in *Marxist*, XXVII 1-2, January-June 2011. https://www.cpim.org/marxist/2011-01-science-society-pabir.pdf.

Chapter 8, 'The Untold Story of the Left in Indian Science', draws on an essay published in *Newsclick* on 17 October 2020, https://www.newsclick.in/The-Untold-Story-Left-Indian-Science, and from the Introduction to the book *Political Journeys in Health, Essays by and for Amit Sengupta*, Prabir Purkayastha et al (eds.), LeftWord, 2020.

Chapter 9, 'Technology in a Postcolonial Setting: Notes from the Subcontinent', draws on a review I wrote of Ghulam Kibria's *Technology Acquisition in Pakistan: Story of a Failed Privileged Class and a Successful Working Class*, City Press, Karachi, 1998, in *Science, Technology and Society*, 6(1), 2001, pp. 241–246. https://doi.org/10.1177%2F097172180100600112.

Section IV: An earlier version of Chapter 10, 'Hindutva, Pythagoras and the Zero', is based on the article 'Hindutva, Mathematics, Pythagoras and Zero', *Newsclick*, 26 November 2015. https://www.newsclick.in/hindutva-mathematics-pythagoras-and-zero.

Chapter 11, 'Building a Nation with a Scientific Vision', draws on the M. Basavapunnaiah Memorial Lecture I delivered on 14 December 2021 in Vijayawada; 'European Nationalism, Khaki Shorts'; published in *Communalism Combat*, Year 10, No. 92, October 2003. https://www.sabrang.com/cc/archive/2003/oct03/forum.html.